Testing Aircraft, Exploring Space

PUBLISHING FOR THE WORLD
125 Years
THE JOHNS HOPKINS UNIVERSITY PRESS

THE JOHNS HOPKINS UNIVERSITY PRESS   BALTIMORE & LONDON

Roger E. Bilstein

Testing Aircraft, Exploring Space

An Illustrated History of NACA and NASA

9 8 7 6 5 4 3 2 1

The Johns Hopkins University Press
2715 North Charles Street
Baltimore, Maryland 21218-4363
www.press.jhu.edu

Library of Congress Cataloging-in-Publication Data

Bilstein, Roger E.
  Testing aircraft, exploring space : an illustrated history of NACA
and NASA / Roger E. Bilstein.
      p.   cm.
Includes bibliographical references and index.
  ISBN 0-8018-7158-1 (hardcover : alk. paper)
    1. United States. National Aeronautics and Space Administra-
tion—History.   2. United States. National Advisory Committee
for Aeronautics—History.   3. Airplanes—United States—Flight
testing—History.   4. Astronautics—United States—History.
5. Outer space—Exploration—History.   I. Title.
  TL521.312 .B58 2003
  629.4'0973—dc21   2002006469

A catalog record for this book is available from the British Library.

To the dedicated men and women of NACA and NASA

Contents

Preface

In 1769 several colonial governments in America funded astronomer John Rittenhouse in a project to observe rare astronomical circumstances involving the planet Venus. The occasion demonstrated an early instance of using public funds in the interests of practically applied science and technology; the project resulted in more accurate celestial charts for improved maritime navigation as well as pathfinding across the trackless wilderness of North America. As the new nation pushed its frontier farther west, other examples of federally funded activities followed the line of settlement to enhance transportation, commerce, and national security. In 1806, for example, Congress earmarked money for the Cumberland Road, running through the Alleghenies into the western lands drained by the Ohio and Mississippi Rivers. After the Civil War, land grants, loans, and other benefits from federal and state sources aided construction of transcontinental railroads that wrought new changes in the American frontier. Supporting the automotive revolution of the twentieth century, federal and state funds built the transcontinental Lincoln Highway during the 1920s and 1930s and launched the Interstate Highway System during the 1950s. The latter also enhanced national security in the cold war era, providing the nation with an alternative transportation system in case hostile forces attacked vulnerable nodes of the railroad network. In its formation as well as its evolution, the National Advisory Committee for Aeronautics (NACA), later the National Aeronautics and Space Administration (NASA), continued a pattern already established by such federal activities.

As a new scientific and technological phenomenon of the twentieth century, the study of flight promised advances in transportation and communication, as well as military applications. Tradition made it a candidate for federal support. More to the point, rapid European progress in the pre–World War I era motivated aeronautical partisans in the United States to advocate action that would help America to catch up. Demonstrations of military air power after the outbreak of the war in Europe made it even more imperative to move toward practical research and development (R&D). In this internationally competitive environment—with overtones of national defense—NACA came into being during 1915. Military concerns continued to occupy NACA researchers during World War II; the cold war era after 1945 spurred additional programs keyed to military aviation. The dramatic launch of the Soviet Union's *Sputnik* in 1957 generated a new sense of technological competition in the context of cold war posturing and led directly to the reorganization of NACA as the National Aeronautics and Space Administration (NASA) in order to explicitly challenge the Soviets in the exploration of space and to beat them in sending a manned mission to the Moon.

All of this produced changes in the ways that NACA and its successor organized operational functions and interacted with the aviation community—and with manufacturers in particular. As early as the 1920s, NACA became more sensitive to the commercial agenda of the aeronautical community; original research in NACA laboratories persevered, but specific projects shaped to specific concerns of the industry became more prevalent. European influences on NACA's activities continued, including major legacies of the World War II era such as jet engines and swept wings. The European connection has been a persistent subtheme in the evolution of many key programs conducted by both NACA and NASA. The period after World War II heralded a fundamental change in many NACA procedures for R&D. In the process of developing the rocket propelled XS-1 supersonic aircraft, NACA assumed the roles of manager and customer, and corporations in the private sector carried out contract work for the agency. The manager-customer role became entrenched during the dramatic growth of NASA's space program, with Marshall Space Flight Center and Johnson Space Center, for example, dispensing billions of dollars to contractors from coast to coast. In Washington, D.C., NASA Headquarters also acquired hundreds of administrative personnel, making it a far different entity from the small, clubby advisory committee that NACA had been.

Given several decades of development, various centers inevitably acquired different looks and different cultures. On the East Coast, Langley Research Center's campus in Virginia included large, strangely lumpy structures to house big wind tunnels and high-speed test facilities. With many administrative buildings dating back to the 1930s and avenues lined by mature trees, Langley seemed to have an older, more settled appearance than other centers and a sedate way of doing business. The Kennedy Space Center, in Florida, featured bigger, blockier buildings to handle towering rocket-launch vehicles and futuristic space shuttles. Tall, girderlike structures punctuated its Atlantic shoreline—launching pads for manned journeys into space. Its sprawling environs—no cozy, hometown flavor here—provided safe buffer zones in case a launch vehicle laden with thousands of gallons of volatile fuel and chemicals exploded. Kennedy's thousands of acres of canals and swamps also offered a protective habitat for subtropical flora and fauna, including alligators and birds. Marshall Space Flight Center, in Alabama, had a compact administrative block with 1960s-era multistory buildings, but most of its real estate was devoted to test areas for huge propellant tanks, brawny rocket engines, and similar hardware for launch vehicles. Its U.S. Army arsenal heritage gave it a distinctive industrial look: it was crisscrossed by railroad tracks to trundle heavy components, and people in hard hats were not uncommon. Johnson Space Center, in Texas, with modernistic architecture and a broad, landscaped greensward in its center, looked not unlike a university campus. To serve its broad management functions, it housed numerous facilities for astronaut training and the mission control center for handling manned missions in space. Not many hard hats here. In California, Dryden Flight Research Center sat surrounded by a barren backdrop of austere, surrealistic mountains. Sited in a high desert region dotted with exotic cacti, Dryden's flight line was usually populated by exotic experimental aircraft. Typical attire for its personnel included dark aviator's glasses and the informal, laid-back attitude affected by test pilots.

In aeronautical research, NASA continued its quest for faster aircraft performance but eventually realized that very high supersonic speeds often compromised other useful qualities of combat planes; new research in the closing decades of the twentieth century often focused on computerized controls, composite structures, new regimes of aerodynamic efficiency, and maneuverability. In the realm of civil aeronautics, the agency achieved commendable advances in air traffic control operations and environment-friendly technologies. The latter seemed to fit the national temper better than exotic manned missions into space. After the United States succeeded in the cold war goal of landing humans on the Moon, congressional and general interest in populating lunar colonies or launching expensive manned missions to distant planets seemed to diminish. Unmanned missions continued to broaden our knowledge of the universe and often generated intense public fascination, but NASA also prudently built on public concern for the global environment with a series of satellite programs and other projects that yielded valuable—if often unsettling—data about global pollution and the state of the world's natural resources.

From its origins as an advisory committee with a skeleton staff in 1915, by the end of the twentieth century, NASA had become a multifaceted federal phenomenon with annual budgets of over $10 billion and some eighteen thousand employees at ten field centers in eight states and the District of Columbia.

There were occasions when some of its organizations fiercely competed against each other for financial resources, especially at the end of the Apollo era when space centers with thousands of NASA engineers suddenly found themselves with time on their hands and no complex manned space missions as national priorities. Overall, NASA's responses to successive peacetime and wartime imperatives, including cold war issues and environmental concerns, clearly signaled the agency's sensitivities to sea changes in the nation's social, cultural, and political life. As the twentieth century ended and the twenty-first began, the end of the cold war, more stringent budgets, and trends toward internationalism in industry left their imprint on NASA as well. Always attuned to developments overseas, NASA continually expanded its international alliances after World War II. Joint space missions with various countries helped build momentum toward the International Space Station in the 1990s. More dramatically, Russia became a major partner in the space station venture. By the start of the twenty-first century, the International Space Station circled the earth with both Russian and American crews aboard. But budgetary constraints—which dictated solicitations for foreign funding—drove planning for the space station and other NASA programs as much as the new international spirit did. Meanwhile, beginning with the Apollo Moon landing program in the 1960s, NASA paid increasing attention to spin-off technologies that benefited non-aerospace enterprises as a means to enlist broader public and political support. This kind of covert public relations campaign marked an unusual new posture for the agency.

By the close of the 1990s, NASA had its back to the wall in financial terms. Moreover, critics took the agency to task for a deteriorating state of affairs in aeronautics. American R&D, the critics charged, had suffered declining budgets for too many years, allowing multinational consortiums in Europe to challenge America's premier leadership in both military and civil aviation. In 2001 the decision by the European Airbus group to proceed with production of a new jumbo jet, outclassing the Boeing 747, was interpreted by many observers as a symbol of American aeronautical decay. Ironically, the same lamentations had been voiced at the beginning of the twentieth century. Once again, NASA faced pressing new challenges at the start of a new century.

Acknowledgments

In 1965 Eugene M. Emme, historian for the National Aeronautics and Space Administration, wrote a brief survey of the agency entitled *Historical Sketch of NASA* (Washington, D.C.: NASA, 1965). It served its purpose as a succinct overview useful for federal personnel, new NASA employees, and inquiries from the general public. Because people were so curious about the nascent space program, the text emphasized astronautics. By 1976 a revision was in order, and it was undertaken by Frank W. Anderson Jr., publications manager of the NASA History Office. With a new title, *Orders of Magnitude: A History of NACA and NASA, 1915–1976,* this version gave a concise summary of NASA's predecessor, the National Advisory Committee for Aeronautics (NACA), although astronautics was still accorded the lion's share of the text. After a second printing, Anderson prepared a revised version, published in 1981, which carried the NASA story up to the threshold of Space Shuttle launches. Meanwhile, Anderson retired from NASA in 1980.

As NASA approached the seventy-fifth anniversary of its 1915 NACA origin, a further updating of *Orders of Magnitude* seemed to be needed. In addition to serving its original audience, the book had been useful as a quick reference and as ancillary reading in various history courses; Anderson's lucid style appealed to many readers, including myself. The opportunity to prepare a third edition (published in 1989) was an honor for me.

Anderson's discussions of astronautics remained essentially intact in the third edition, albeit blended into substantial additions that recognized NACA's seminal contributions to aeronautical progress. Three new chapters at the beginning of the book carried the NACA story up to the origins of NASA in 1958. I also revised all four older chapters on the space program to include discussion of aviation, in order to chronicle the continuing evolution of aeronautical research during the Apollo era. I wrote two final chapters describing developments in aeronautics and astronautics up to 1990, added a short bibliographical essay, and replaced a number of photos.

With the current edition (the fourth), the NASA History Office submitted the manuscript to the Johns Hopkins University Press. The book also received a new title. One of the strongest virtues of prior versions has been their brevity, and I have made a special effort to keep this book's size under control as well. Clearly, neither this narrative history nor its precursors were ever intended as a definitive or interpretive study of NACA and NASA. I hope that this succinct coverage will maintain the book's usefulness and appeal as an accessible, concise overview of the agency. Accordingly, all chapters have been pruned (some more than others), and two short chapters of the prior edition have been folded together. Although I have added two new chapters that cover recent developments, the overall length of the main text is close to that of its predecessor. New features of the fourth edition include a chronology and a list of abbreviations.

I continue to owe a debt of thanks to the individuals noted in the third edition. For this new iteration, Audrey Schwartz-Rivers, formerly of Johnson Space Center Public Affairs, supplied indispensable information and commentary. Discussions with Sascha Tarrant, a graduate student from the University of Houston, Clear Lake, sharpened my focus about the chronology. Anonymous referees helped strengthen the narrative and resolve several incongruities. At the university's Research Support Office, I wish to thank Robert Hodgin, Director, for facilitating this effort; special appreciation goes to Kathy Dupree, who sorted out a variety of contractual issues. Agatha Whitney, our suite secretary, patiently converted early handwritten drafts into the requisite computerized formats. At the Johns Hopkins University Press, plaudits go to Robert J. Brugger and to his editorial assistant, Melody Herr, for ongoing encouragement, assistance, and patience. My family once again weathered the process of historical composition. My wife, Linda, endured with good humor the sporadic clutter of books and notes throughout our house, while family members tolerated a new roster of aerospace anecdotes during their visits home.

My initial experience as a NASA contract historian occurred in 1969, while Gene Emme still served as NASA's first official Headquarters historian. Over the past thirty-plus years, I have had the good fortune to work for all three subsequent managers of the NASA Headquarters History Office: Monte Wright, Sylvia Kraemer, and Roger Launius. All of them have offered welcome encouragement and sound advice. Exchanges with Roger Launius about recent trends were especially informative.

In the process of defining the coverage and topics in these surveys and other history projects for NASA during the course of three decades, I have always had the professional freedom to establish my own agenda. As in the past, any shortcomings and errors are mine alone.

Abbreviations

ABMA	Army Ballistic Missile Agency
DoD	Department of Defense
EVA	Extravehicular activity
GALCIT	Guggenheim Aeronautical Laboratory, California Institute of Technology
ICBM	Intercontinental ballistic missile
ISS	International Space Station
JPL	Jet Propulsion Laboratory
KSC	Kennedy Space Center
LM	lunar module
Mach	unit of speed, equivalent to the speed of sound
MSFC	Marshall Space Flight Center
NACA	National Advisory Committee for Aeronautics
NASA	National Aeronautics and Space Administration
R&D	research and development
STOL	short takeoff and landing
STS	Space Transportation System
VfR	Verein für Raumschiffahrt (Society for Space Travel)

Testing Aircraft, Exploring Space

Chapter 1

Foundations for Flight, 1915–1930

During World War I the American aviation community began to realize that European aviation had outstripped progress in the United States. With the added threat of rapid wartime development overseas, the United States hastily organized the National Advisory Committee for Aeronautics (NACA). Within its first fifteen years, that agency erected several impressive wind tunnels, shrewdly kept track of advanced European aeronautics (even hiring experts from Europe), and soon established itself as one of the world's leading centers in flight research. In the process of producing data for American civil aviation, NACA also played a key role in advancing military aeronautics. During the post–World War I era, individual American and European experimenters also took the first steps toward practical rocketry.

Planning for Flight Research

In 1915, congressional legislation created an Advisory Committee for Aeronautics. The prefix "National" soon became customary and was officially adopted, and the familiar acronym NACA emerged as a widely recognized term throughout the aeronautics community in America. The genesis of the committee occurred at a time of accelerating cultural and technological change. Only the year before, Robert Goddard had begun experiments in rocketry and the Panama Canal had opened. Amid the gathering whirlwind of World War I, social change and technological transformation persisted. During 1915,

NACA's first year, Albert Einstein postulated his general theory of relativity and authorities jailed Margaret Sanger as the author of *Family Limitation*, the first popular book on birth control. Frederick Winslow Taylor, father of "Scientific Management," died, and disciples like Henry Ford diligently applied his ideas to achieve prodigies of production. Ford produced his one-millionth automobile the same year. In 1915 Alexander Graham Bell made the first transcontinental telephone call, from New York to San Francisco, with his trusted colleague Dr. Thomas A. Watson on the other end of the line. Motion pictures began to reshape American entertainment habits, and New Orleans jazz began to make its indelible imprint on American music. At Sheepshead Bay, New York, racing drivers set a new speed record for automobiles, 102.6 MPH, a figure that many fliers of the era would have been happy to match.

American flying not only lagged behind automotive progress but also lagged behind European aviation. Many aviation enthusiasts in the United States, the home of the Wright brothers, found this particularly galling. True, Orville and Wilbur Wright had benefited from the work of European pioneers such as Otto Lilienthal in Germany and Percy Pilcher in Great Britain. But it was the Wrights who made the first powered, controlled flight in an airplane on 17 December 1903, on a lonely stretch of beach near Kitty Hawk, North Carolina. Ironically, this feat was widely ignored or misinterpreted by the American press for many years, until 1908, when Orville made trial flights for the War Department and Wilbur's flights overseas enthralled Europe. Impressed by the Wrights, the Europeans nonetheless had already begun a rapid development of aviation, and their growing record of achievements underscored the lack of organized research in the United States.

Sentiment for some sort of center of aeronautical research had been building for several years. At the inaugural meeting of the American Aeronautical Society, in 1911, some of its members discussed a national laboratory with federal patronage. The Smithsonian Institution seemed a likely prospect, based on its prestige and the legacy of Samuel Pierpont Langley, who was its former secretary, an aviation pioneer, and a frustrated competitor of the Wrights. But Langley's dusty equipment rested where it had been abandoned in his lab behind the Smithsonian "castle" on the Mall, and the defensive attitude of the Smithsonian about the first manned flight presented a prickly barrier. The American Aeronautical Society's dreams became entangled and frustrated by continued in-fighting among other organizations that were just beginning to see aviation as a promising research frontier, including universities like the Massachusetts Institute of Technology and government agencies like the U.S. Navy and the National Bureau of Standards.

The difficulties of defining a research facility were compounded by the ambivalent attitude of the American public toward the airplane. While some saw it as a mechanical triumph with a significant future, others saw it as a mechanical fad, and a dangerous one at that. If anything, the antics of the "bird-men" and the "aviatrixes" of the era tended to underscore the foolhardiness of aviation and airplanes. Fliers might set a record in one month and fatally crash the next. Calbraith P. Rodgers managed to make the first flight from the Atlantic to the Pacific coast in 1911 (in forty-nine days, with nineteen crashes and innumerable stops), but he died in a crack-up just four months later. Harriet Quimby, the attractive and chic American aviatrix (she flew wearing a specially designed, plum-colored satin tunic), made headlines as the first woman to fly across the English Channel in 1912. After returning to America, she died in a crash off the Boston coast within three months.

Fatalities also occurred in Europe, but the Europeans also took a different view of aviation as a technological phenomenon. Governments, as well as industrial firms, tended to be more supportive of what might be called "applied research." As early as 1909, the internationally known British physicist Lord Rayleigh assumed leadership of Britain's own Advisory Committee for Aeronautics; in Germany, Ludwig Prandtl and others were beginning the sort of investigations that soon made the University of Göttingen a center of theoretical aerodynamics. Additional programs gathered momentum in France and elsewhere on the Continent. Similar progress in the United States barely limped along. Aware of European activity, Charles D. Walcott, secretary of the Smithsonian Institution, scraped together funds to dispatch two Americans on a fact-finding tour overseas. Dr. Albert F. Zahm taught physics and experimented in aeronautics at Catholic University in Washington, D.C.; Dr. Jerome C. Hunsaker, a graduate of the Massachusetts Institute of Technology, was developing a curriculum in aeronautical engineering at the institute. Zahm and Hunsaker's report, issued in 1914, emphasized the exasperating disparity between European progress and American inertia. The visit also established European contacts that later proved valuable to NACA.

The outbreak of war in Europe in 1914 served as a catalyst for the creation of an American agency. The use of German dirigibles for long-range bombing of British cities and the rapid evolution of airplanes for reconnaissance and pursuit underscored the shortcomings of American aviation. Against this background, Walcott pushed for legislative action to provide for aeronautical research that would allow the United States to match progress overseas. Walcott received support from Progressive leaders in the country, who felt that government agencies for research were consistent with Progressive ideals such as

Pre–World War I aviation technology. Military personnel flight-tested a Wright Model A biplane during trials at Fort Myer, Virginia, in September 1903.

scientific inquiry and technological progress. By the spring of 1915, the drive for an aeronautical research organization finally succeeded.

The enabling legislation for NACA slipped through almost unnoticed as a rider attached to the Naval Appropriation Bill, on 3 March 1915. It entailed a traditional example of American political compromise.

As before, the move had been prompted by the Smithsonian. The legislation did not call for a national laboratory, since President Wilson apparently felt that such a move, taken during wartime conditions in Europe, might compromise America's formal commitment to strict nonintervention and neutrality. Although supported by the Smithsonian, the proposal emphasized a collective responsibility through a committee that would coordinate work already under way. The committee comprised an unpaid panel of 12 people, including 2 members from the War Department, 2 from the Navy Department, 1 each from the Smithsonian, the Weather Bureau, and the Bureau of Standards, and 5 more members acquainted with aeronautics. Despite concerns about flouting the conventions of neutrality, the proposal's proponents in Congress tacked it on as a rider to the naval appropriation bill as a ploy to clear the way for quick endorsement.

In shaping NACA's charter, Hunsaker and his colleagues clearly borrowed language from the British Advisory Committee for Aeronautics, including the precise wording of NACA's well-known mandate to pursue "the scientific study of the problems of flight, with a view to their practical solution." The British organization also served as the model for NACA's committee structure and preliminary notations for potential aeronautical investigations. In myriad ways, European legacies not only marked NACA's origins but also punctuated its subsequent history.

For fiscal 1915 the fledgling organization received a budget of five thousand dollars, an annual appropriation that remained constant for the next five years. Not much even by standards of that time, but it must be remembered that NACA was an advisory committee only. Once NACA isolated a problem, the study and solution of the problem generally became the task of a government agency or a university laboratory, often on an ad hoc basis within limited funding. The main committee of twelve members met semiannually in Washington; an Executive Committee of seven members, characteristically chosen from the main committee members living in the Washington area, supervised NACA's activities and kept track of aeronautical problems to be considered for action. Although a clubby arrangement, it seemed to work.

In the wartime environment, NACA quickly found itself quite busy. It evaluated aeronautical queries from the army and conducted experiments at the navy yard; the Bureau of Standards ran engine tests; Stanford University ran propeller tests. But NACA's role as mediator in the rancorous and complex squabble between Glenn Curtiss and the Wright-Martin Company represented its greatest wartime success. The dispute involved the technique for lateral control of aircraft in flight and dated back to prewar controversies involving Langley and Curtiss versus the Wrights. Once settled, the resultant cross-licensing agreement consolidated patent rights and cleared the way for volume production of aircraft during the war as well as during the postwar era.

The authors of NACA's charter had left open the possibility of an independent laboratory. Although several facilities for military research continued to function, NACA pointed out in its annual report for 1915 that civil aviation research would be in order when the Great War ended. And so, even before the war's conclusion, plans were afoot to acquire a laboratory. The best option seemed to be collaboration in the development of a new U.S. Army airfield, across the river from Norfolk, Virginia. The military facility was named after Samuel Pierpont Langley, former secretary of the Smithsonian; the NACA facility was named the Langley Memorial Aeronautical Laboratory, soon shortened to the familiar, cryptic "Langley."

Construction of the airfield got under way in 1917, hampered by the confusion following America's declaration of war on Germany and by the wet weather and marshy terrain of the Virginia tidewater region. The throng of workers included an aspiring young writer named Thomas Wolfe. In his autobiographical novel, *Look Homeward Angel* (1929), Wolfe's main character found a job at Langley as a horse-mounted construction supervisor paid eighty dollars per month. He directed gangs striving to create a level airfield, pushing the earth "and filling interminably, ceaselessly, like the weary and fruitless labor of a nightmare, the marshy earth-craters, which drank their shoveled toil without end."

But eventually it did end; formal dedication took place on 11 June 1920. Although the army, under wartime pressures, had already relocated its own research center to McCook Field, near Dayton, Ohio, Langley Field remained a large base, and military influence continued to be strong. The inaugural ceremonies included various aerial exhibitions and a flyover of a large formation of planes led by the dashing brigadier general William "Billy" Mitchell. Visitors found that NACA's corner of Langley Field occupied a comparatively modest site: an atmospheric wind tunnel, a dynamometer lab, an administration building, and a small warehouse. The staff totaled only eleven people—plenty of room to grow.

The Postwar Era

The management of NACA and Langley, with its small contingent for so many years, remained personal, straightforward, and more or less informal. In Washington a full-time executive secretary was named: John F. Victory, NACA's first employee, hired in 1915. George W. Lewis, hired in 1919, became director of research but remained in Washington, where he could palaver with politicians and joust with other bureaucrats. He spent long, productive hours in the corridors of the Army-Navy Club and the Cosmos Club. Meanwhile, the close-knit staff down at Langley operated on a more democratic basis. In the lunchroom junior staff, senior staff, and technicians dined together, where a free exchange of views continued over coffee and dessert. For years Langley managed to attract the brightest young aeronautical engineers in the country, because they knew that their training would continue to expand by close and comradely contact with many senior NACA engineers on the cutting edge of research.

Engineers came to Langley from all over the country. Early employees often had degrees in civil or mechanical engineering, since so few universities

Langley Laboratory's first wind tunnel, a replica of a ten-year-old British design, became operational in June 1920.

offered a degree in aeronautical engineering alone. By the end of the 1920s, this had begun to change. From a handful of prewar courses dealing with aeronautical engineering, universities like the Massachusetts Institute of Technology evolved a plan of professional course work leading to both undergraduate and graduate degrees in the subject. The Daniel Guggenheim Fund for the Promotion of Aeronautics provided money for similar programs at several other schools. In 1929 a survey by an aviation magazine reported that fourteen hundred aeronautical engineering students were enrolled in more than a dozen schools across the United States. The California Institute of Technology became a major beneficiary of the Guggenheim Fund's foresight. Although America possessed the facilities to train engineers, and NACA offered superb facilities for practical research, the country lacked a nerve center for advanced studies in theoretical aerodynamics. Germany led the world in this respect until the Guggenheim Fund lured the brilliant young scientist Theodore von Karman to the United States. Von Karman accepted a Caltech offer in 1929 and occupied his new post the following year. Within the decade, not only did Caltech's research projects enrich the field of aerodynamic theory, but also its graduates became leading aeronautical educators in colleges and universities across the nation. The Guggenheim Fund's largesse constituted a tremendous stimulus to aeronautical engineering and research and to the dozens of other aeronau-

A NACA team conducts research using the variable density tunnel at the Langley Memorial Aeronautical Laboratory in 1929.

tical projects that it supported. Between 1926 and 1930, this personal philanthropy disbursed $3 million for a variety of fundamental research and experimental programs, including flight safety and instrument flying, that profoundly influenced the growth of American aviation.

Although the Langley organization became more formalized over time, the young organization allowed maximum opportunity for individual initiative. The agency followed a regular procedure for instituting a "Research Authorization," but promising ideas could be pursued without formal approval. The NACA hierarchy in Washington and at Langley accepted this sort of "bootlegged" work as long as it did not appear to be too exotic, because it often proved to be as productive as formal programs, and it kept the Langley staff moving out in front of the conventional frontier. The system also worked because the Langley staff remained small: about one hundred in 1925. Creativity had its place, but senior engineers quickly spotted outlandish projects. The sources for formal Research Authorizations were many and varied, often reflected by the catholic makeup of NACA's main committee, which drew from both military services, other government agencies, universities, and the aviation community. Ideas also came from Lewis's forays into Washington corridors of influence as well as from sources overseas. Edward Pearson Warner, serving as Langley's chief physicist, was packed off to Europe in 1920 to get a sense of postwar trends among major overseas countries; later NACA set up a permanent observation post in Paris, where John J. Ide kept an eye on European activities and relayed pertinent information up to World War II.

But research depended on facilities. At Langley, NACA technicians turned their attention to a new wind tunnel. It was not large, designed to have a test section about five feet in diameter, but it could be configured to produce speeds of 120 MPH in the test section, making it one of the best facilities in the world. Still, there were inherent drawbacks. With no firsthand experience, NACA planners built a conventional, open-circuit tunnel based on a design used at the British National Physical Laboratory. Electric fans drew air directly from the atmosphere, introducing a rapid airflow into the test section; the air exited directly back into the building that housed the tunnel. Test results often tended to be variable, because of the humidity and atmospheric turbulence that resulted within the walls of the test laboratory. At the University of Göttingen in Germany, the famous physicist Ludwig Prandtl and his staff had already built a closed-circuit, return-flow tunnel in 1908. Among other things, the closed-circuit design required less power, boasted a more uniform airflow, and permitted pressurization as well as humidity control.

The NACA engineers at Langley knew how to scale up data from the small models tested in their sea-level, open-circuit tunnels, but they soon realized that their estimates were often wide of the mark. For significant research, NACA experimenters needed facilities like the tunnels in Göttingen. They also needed a researcher with experience in the design and operation of these more exotic tunnels. Both requirements were met in the person of Max Munk.

Munk had been one of Prandtl's brightest lights at Göttingen. During World War I many of Munk's experiments in Germany were instantaneously tagged as military secrets (though they usually appeared in England, completely translated, within days of his completing them). After the war Prandtl contacted his prewar acquaintance Jerome Hunsaker with the news that Munk wanted to settle in America. For Munk to enter the United States in 1920, President Woodrow Wilson had to sign two special orders: one to get him into America so soon after the war, and one permitting him to hold a government job. In the spring of 1921, construction of a pressurized, or variable density, tunnel began at Langley. The goal was to keep using models in the tunnel but to conduct the tests in a sealed, airtight chamber where the air would be compressed "to the same extent as the model being tested." In other words, if a one-twentieth-scale model was being tested in the variable density tunnel, then researchers would increase the density of air in the tunnels to twenty atmospheres. Results could be expressed in a numerical scale known as the Reynolds number. The tunnel began operations in 1922 and proved highly successful in the theory of airfoils. As one Langley historian wrote, "Langley's VDT [variable density tunnel] had established itself as the primary source for aerodynamic data at high Reynolds numbers in the United States, if not in the world." Munk's tenure at NACA was a stormy one. He was brilliant, erratic, and an autocrat. After many confrontations with various bureaucrats and Langley engineers, Munk resigned from NACA in 1929. But his style of imaginative research and sophisticated wind tunnel experimentation bequeathed a significant legacy to the young agency.

The variable density tunnel, using scale models, involved only one avenue of aeronautical investigation. In parallel, NACA ran a program of full-scale flight tests that also yielded early dividends. In the process NACA helped establish a body of requisite guidelines and procedures for flight testing. One problem involved instrumentation—proper equipment for acquiring accurate data on full-scale aircraft during actual flight that could correlate with data obtained in wind tunnels. In one early project, wind tunnel data for a model of the Curtiss JN-4 ("Jenny") was compared to information derived from an instrumented Jenny put through a series of flight tests, in order to investigate lift and drag. By comparing data, the reliability of wind tunnel information could be judged more rigorously. The tests of the 100 MPH JN-4 marked the beginning of carefully planned and instrumented experimental flights, which became a hallmark of NACA and NASA from subsonic through supersonic flight. The early JN-4 flights also uncovered another aspect of flight testing to be addressed—the need for specially trained test pilots. Langley also pioneered in the concept of training fliers as test pilot–engineers.

By 1922 several different kinds of aircraft were under test at Langley. Three workhorse planes were Curtiss JN-4H Jennies, used for a series of takeoff and landing and performance measurements, in order to establish new design parameters. Military investigations also began during these early years, when the Navy Bureau of Aeronautics came to NACA for a comparative study of airplanes in terms of stability, controllability, and maneuverability. Along with a Vought VE-7 from the navy, Langley pilots obtained a Thomas-Morse MB-3 from the army and two foreign models: a British SE-5A (one of the Royal Air Force's principal fighters of World War I) and a German Fokker D-VII (the main source of references to the "Fokker scourge" during the war). Evaluating frontline aircraft from foreign as well as American air forces inaugurated a practice that persisted through the NASA era as well. Other investigations dur-

A Langley researcher ponders the future, in mid-1927, of the Sperry M-1 Messenger, the first full-scale airplane tested in the propeller research tunnel.

ing the mid-1920s involved further work for the navy, to ascertain accurate data on stall, takeoff, and landing speeds of a specific aircraft. The army turned up with a similar request for studies of these and other qualities for most of the aircraft in the Air Service inventory at that time.

The progressive experience in flight test work, including a variety of instrumentation required to register the data, contributed to studies of pressure distribution along wing surfaces, a major effort during the 1920s. Beginning with measurements during steady flight, test pilots and instrumentation experts devised techniques to study pressure distribution during accelerated flight and in maneuvers, accumulating invaluable design data where none had existed before. Steady improvement in instrumentation permitted pressure distribution surveys to be wound up in one day, rather than making a prolonged series of flights lasting as long as two months. By 1925 Langley had nineteen aircraft dedicated to various test operations. Ground testing had expanded to include a new engine research laboratory, in which engineers had begun work on supercharging engines for high altitude bombers and on a means of boosting power for interceptors in order to give them a high rate of climb—the sort of investigative work that paid dividends later in World War II.

The Tunnels Pay Off

In the meantime, the variable density tunnel began to pay further dividends in the form of airfoil research. During the late 1920s and into the 1930s, NACA developed a series of thoroughly tested airfoils and, for each one, devised a four-digit numerical designation that indicated the airfoil section's critical geometric properties. By 1929 Langley had developed this system to the point where the numbering system was complemented by an airfoil cross section, and the complete catalog of seventy-eight airfoils appeared in NACA's annual report for 1933. Engineers could quickly see the peculiarities of each airfoil shape, and the numerical designator ("NACA 2415," for instance) specified camber lines, maximum thickness, and special nose features. These figures and shapes gave engineers the information they needed to select specific airfoils for desired performance characteristics of specific aircraft.

During the late 1920s, NACA also announced a major innovation that resulted in the agency's first Robert J. Collier Trophy, presented annually by the National Aeronautic Association for the year's most outstanding contribution to American aviation. In 1929 the Collier trophy went to NACA for the design of a low-drag cowling.

Most American planes of the postwar decade mounted air-cooled radial engines, with the cylinders exposed to the airstream to maximize cooling. But the exposed cylinders also caused high drag. Because of this, the U.S. Army had adopted several aircraft with liquid-cooled engines, in which the cylinders were arranged in a line parallel to the crankshaft. This reduced the frontal area of the aircraft and also permitted an aerodynamically contoured covering, or nacelle, to be placed over the nose of the plane. But the liquid-cooled designs carried weight penalties in terms of the many cooling chambers around the cylinders, the gallons of coolant, the pumps, and the radiator. The U.S. Navy decided not to use such a design because the added maintenance requirements cut into the limited space aboard aircraft carriers. Moreover, the jarring contact of airplanes with carrier decks created all sorts of cracked joints and leaks in liquid-cooled engines. Air-cooled radial engines simplified this issue, although their inherent drag meant reduced performance. In 1926 the navy's Bureau of Aeronautics approached NACA to see if a circular cowling could be devised in such a way as to reduce the drag of exposed cylinders without creating too much of a cooling problem.

While significant work on cowled radial engines proceeded elsewhere, particularly in Great Britain, investigations at Langley soon provided a breakthrough. American aerodynamicists at this time had the advantage of a new propeller research tunnel that was completed at Langley in 1927. The tunnel's diameter of twenty feet made it possible to run tests on a full-sized airplane. Following hundreds of tests, a NACA technical note by Fred E. Weick in November 1928 announced convincing results. At the same time, Langley acquired a Curtiss Hawk AT-5A biplane fighter from the Air Service and fitted a cowling around its blunt radial engine. The results were exhilarating. With little additional weight, the Hawk's speed jumped from 118 to 137 MPH, an increase of 16 percent. The virtues of NACA cowling received public acclaim the next year, when Frank Hawks, a highly publicized stunt flier and air racer, added NACA cowling to a Lockheed Air Express monoplane and racked up a new Los Angeles–New York nonstop record of eighteen hours and thirteen minutes. The cowling had raised the plane's speed from 157 to 177 MPH. After the flight, Lockheed Aircraft sent a telegram to the NACA committee: "Record impossible without new cowling. All credit due NACA for painstaking and accurate research." NACA estimated that using the cowling would result in savings to the industry of over $5 million—more than all the money appropriated for NACA from its inception through 1928.

After fifteen years the sophistication of NACA's research had dramatically changed. And so had the sophistication of aviation. After a fitful start in 1918, the U.S. government's airmail service gradually forged day-and-night transcontinental routes across America by 1924. The service saved as much as two days in delivering coast-to-coast mail, accelerating the tempo of the business civilization and saving millions of dollars. In 1925 the government began to contract for service with privately owned companies, a change that marked the beginning of the airline industry. By the end of the decade, the private companies advertised to fly passengers as well as mail, and Pan American Airways launched international services between Florida and Cuba, as well as between Texas and Central America. Following the Air Commerce Act of 1926, lighted airways were improved, radio communications progressed, and federal inspectors enforced standards for pilot proficiency as well as aircraft design and construction. By the time Charles Lindbergh made his solo flight from New York to Paris in 1927, an aeronautical infrastructure was already in place. The "Lindbergh Boom" that followed his striking achievement could not have been sustained without the important progress of the previous years.

NACA helped spur much of this development through its refinement of wing design and investigations of various aerodynamic phenomena. The agency also benefited from overall aviation progress during this era, sharing the increased aviation budgets for civil programs under the Air Commerce Act and for the expansion of U.S. Army and U.S. Navy aviation. The Army Air

Robert H. Goddard, with the first successful liquid-fueled chemical rocket, launched 16 March 1926.

Service received more autonomy in 1926, when it became the Air Corps. During the 1920s the army's air arm began to develop a doctrine, standardize its training, and pursue advanced research, often in cooperation with NACA. In the development of equipment, the Air Service undertook projects for modern fighters and strategic bombers to come. The U.S. Navy experienced similar organizational changes and began the construction and operational evaluation of aircraft carriers, such as the *Langley*, the *Lexington*, and the *Saratoga*.

Collectively, the progress of civilian aviation, military aviation, and aeronautical research set the stage for the aeronautical revolution that began in the 1930s. The design characteristics of the 1920s—fabric-covered biplanes with radial engines—gave way to the truly sophisticated airplanes of the 1930s with streamlined shapes, metal construction, retractable landing gear, and high performance. Although the national economy sagged during the Great Depression of the 1930s, the aviation industry reached new levels of excellence.

Early Rocketry

There were some areas of flight technology, such as rocketry, in which NACA did not become involved. Nevertheless, when NACA was transformed into NASA in 1958, the new space agency could reach back into some forty years of American and European writing and research on rocketry and the possibilities of spaceflight. During the 1920s the subject of spaceflight often seemed to be the province of cranks and science fiction writers. But visionary researchers in the United States, as well as in Great Britain, Germany, Russia, and elsewhere, were taking the first hesitant steps toward actual space travel. In America Robert Hutchings Goddard is remembered as one of the foremost pioneers.

After completing a doctorate in physics at Clark University in 1911, Goddard joined its faculty. During his physics lectures, he sometimes startled students by outlining various ways of reaching the Moon. Despite the students' skepticism, Goddard was basing his projections on very real advances in metallurgy, thermodynamics, navigational theory, and control techniques. Twentieth-century technology had begun to make rocketry and spaceflight feasible. Goddard fabricated a series of test rockets, and in 1920 he wrote a classic monograph, *A Method of Attaining Extreme Altitudes,* published by the Smithsonian. In it, he described how a small rocket could soar from Earth to the Moon and detonate a payload of flash powder on impact, so that observers using large telescopes on Earth could verify the rocket's arrival on the lunar surface. Caustic news stories about rocketry and lunacy caused Goddard, a shy individual, to shun publicity during the remainder of his life.

Goddard continued to experiment with liquid propellant rockets, igniting them in a field on his Aunt Effie's farm, where their piercing screeches disturbed the neighbor's livestock. Eventually, on 16 March 1926, one of Goddard's devices lifted off to make the first successful flight of a liquid propellant rocket. It was hardly an earth-shaking demonstration—a flight of 2.5 seconds that carried the rocket to an altitude of forty-one feet. A small but significant step toward future progress. Continued work caught the attention of Charles Lindbergh, who persuaded the Guggenheim Fund to support Goddard's research. By the 1930s Goddard set up shop at a desert site near Roswell, New Mexico, where he and a small group of assistants developed liquid propellant rockets of increasing size and complexity. Unfortunately, Goddard's reticence meant that he labored in isolation; other experimental groups knew little of his activities. "His own penchant for secrecy set him apart from the mainstream," wrote historian Frank Winter. "As a result, Goddard's monumental advances in liquid-fuel technology were largely unknown until as late as 1936 when his second Smithsonian report, Liquid Propellant Rocket Development appeared." In the meantime, researchers in Germany began work that eventually had an impact on the American space program.

Rocket enthusiasts in Germany took inspiration from the same science fiction (Jules Verne and others) that had motivated Goddard and took advantage of advances in metallurgy and chemistry. They also took another important step, establishing an organization that facilitated the exchange of information and accelerated the rate of experimentation. In 1927 the Verein für Raumschiffahrt (VfR) was founded by Hermann Oberth and others. A year later the VfR collaborated with producers of a science fiction film on space travel, *The Girl in the Moon*. The script included the now-famous countdown sequence before ignition and liftoff. For publicity, the VfR hoped to build and launch a small rocket. The rocket project fizzled, but on the design team was an eager eighteen-year-old student named Wernher von Braun, whose enthusiasm for spaceflight never waned.

In Russia, Konstantin Tsiolkovsky left a legacy of significant writing in the field of rocketry. Although Tsiolkovsky did not construct any working rockets, his numerous essays and books helped point the way to practical and successful space travel. Tsiolkovsky spent most of his life as an unknown mathematics teacher in the Russian provinces, where he made some pioneering studies in liquid chemical rocket concepts and recommended liquid oxygen and liquid hydrogen as the best propellants. In the 1920s Tsiolkovsky analyzed and mathematically formulated the technique of staging vehicles to reach escape velocities from Earth. Rocket societies were organized as early as 1924 in the Soviet Union, but the barriers of distance and politics limited interchange between these groups and their Western counterparts. In 1931 the Group for the Study of Reaction Motion, known by its Russian acronym GIRD, was organized, with primary research centers in Moscow and Leningrad. GIRD's work resulted in the Soviet Union's first liquid fuel rocket launch in 1933. Although GIRD stimulated considerable activity in the Soviet Union, including conferences, periodicals, and hardware development, military influences became increasingly dominant. The devastating purges of the 1930s seem to have decimated the astronautic leadership in the Soviet Union; the rapid recovery of Soviet activity in the postwar era was therefore all the more remarkable.

Astronautics became professionalized, much as aeronautics had. The term *astronautics* also became more commonplace. That designation grew out of a dinner meeting in Paris in 1927. A Belgian science fiction author, J. J. Rosny, came up with the word, which was then popularized by the French writer and experimenter Robert Esnault-Pelterie, whose best-known book, *L'Astronautique*, appeared in 1930. With a body of literature, evolving technology, active professionals, and an identity, astronautics—like aeronautics—was poised for rapid growth.

Chapter 2

Aeronautics in Peace and War, 1930–1945

In the view of many NACA engineers, remarkable aeronautical progress occurred during the agency's first fifteen years. The next fifteen years, from 1930 to 1945, seemed even more remarkable: streamlined aircraft became commonplace, World War II spawned an impressive variety of modern combat planes, and rocketry became an awesome force in twentieth-century warfare. The administrative and operational nature of NACA experienced subtle but fundamental influences, as the agency added new research centers and interacted more intimately with military initiatives and aviation industry requirements.

Modern Aeronautics

The propeller research tunnel at Langley continued to yield significant information that resulted in equally significant design refinements in the new generation of airplanes. One of the most obvious issues had to do with fixed landing gear and the drag it created. As a means to increase speed, retractable landing gear was not unknown, since this approach had been tried on various airplanes before and after World War I. But retractable gear required additional equipment for raising and lowering and appeared to lack the ruggedness and reliability of conventional, fixed gear. Fixed gear, however, was thought to be a major drag factor, although nobody had accurately assessed the aerodynamic liability. NACA engineers set up a series of tests using the propeller research tunnel to get an accurate measure of the fixed gear's drag on a Sperry Messenger. The results astonished everyone, even NACA staff. According to the figures, fixed gear created nearly 40 percent of the total drag acting on the plane. This eye-opening news, a dramatic demonstration of the performance penalty incurred by using fixed gear, prompted rapid development of retractable gear for a wide variety of airplanes. NACA's tests thus played a large role in the evolution of modern, retractable-geared aircraft.

Additional trends pointed the way to sleeker airplanes, which emerged by the end of the 1930s. Trimotored airliners, such as the Fokkers, the Fords, and the Boeings, had become standard equipment in the United States and elsewhere during the late 1920s. They could not easily be redesigned to mount retractable gear, but the trio of big, blunt radial engines that powered them could be shrouded with the new NACA cowling to give them much improved performance. Engineers at Langley took a Fokker trimotor powered by three Wright J-5 Whirlwind engines and fitted it with cowlings. Their confident expectations of a sudden enhancement of performance were dashed, and the engineers were baffled. They began to wonder whether the installation of engines had something to do with it. So as not to encumber the wing, the original designers had placed the engines on struts beneath the wing (or, in the case of biplanes like the Boeing 80, between the wings). After getting the big Fokker set up in the propeller research tunnel, Langley engineers ran a series of tests that conclusively changed the looks of future multi-engine transports. They discovered that the best position for the engines was neither above nor below the wing, but mounted as part of its structure—situated ahead of the wing, with the engine nacelle faired into the wing's leading edge.

This was the sort of information that also contributed to the development of the modern airliners of the decade. Conventional wisdom in the past had dictated that wings should be mounted high on the fuselage, permitting the engines to be slung underneath with clearance for the propeller arc. This required complex struts (creating drag) and led to the use of awkward, long-legged fixed gear (creating even more drag). If the engines were mounted in the wing's leading edge, the wing could be positioned on the lower part of the fuselage, which meant that the landing gear was now short-legged and less awkward—in fact, retractable. Partially as a result of NACA research, low-winged monoplanes with retractable gear soon replaced the high-winged airliners and many other aircraft of similar design.

The propeller research tunnel at Langley had obviously been a profitable facility, although it had limitations for thorough testing of full-sized aircraft. In 1931, when the full-scale tunnel was officially dedicated, Langley engineers

The Vought O3U-1 became the first complete airplane to be tested in the full-scale wind tunnel, finished in 1931. NASA Photo 75-H-238.

used it to launch a new round of evaluations, which, while sometimes less dramatic than the cowling research, unquestionably added new dimensions to the science of aerodynamics. Its impressive statistics marked the beginning of test facilities of heroic proportions.

Nonetheless, the full-scale tunnel did not overshadow other Langley test facilities. There were those who felt that the shortcomings of the variable density tunnel, with its acknowledged drawbacks in turbulence, would soon be eclipsed by the huge full-scale tunnel. With partisans on both sides, friction between personnel from the variable density tunnel and the full-scale tunnel became legendary. In time, both established relevant niches in the scheme of things. Meanwhile, the variable density tunnel played a key role in many projects, and its personnel made a singular contribution to the theory of the laminar flow wing.

The variable density tunnel could test many more aircraft designs, which could be built as scale models, but the turbulence issue continued to dog research findings. In the process of studying this issue, researchers took a closer look at flow phenomena, especially the "boundary layer," where so many problems seemed to crop up. The boundary layer was known to be a thin stratum of air only a few thousandths of an inch from the contour of the airfoil. Within it, the movement of air particles changed from a smooth laminar flow at the leading edge to a more turbulent state toward the trailing edge. In the process, drag increased. After observing tests in a smoke tunnel and evaluating other data, aerodynamicists concluded that the prime culprits in disrupting laminar flow were traceable to irregularity in the wing's surface (rivet heads and other rough areas) and to pressure distribution over the wing's surface.

Eastman Jacobs, head of the variable density tunnel section, came up with various formulas to allow for the tunnel's turbulence in evaluating models and pushed for a larger, improved tunnel. He also championed a systematic experimental approach in airfoil development.

Jacobs was often challenged by a Norwegian emigrant, Theodor Theodorsen, of the Physical Research Division. Theodorsen, steeped in mathematical research, was a strong proponent of airfoil investigation by theoretical study. His opposition to Jacobs' proposal for an improved variable density tunnel and his insistence that, instead, Langley personnel needed more mathematical skills and theoretical concepts sharpened the debate between experimentalists and theorists within NACA. Jacobs, in fact, kept abreast of current theories, and he eventually fashioned a theoretical approach, backed up by his trademark experimental style, that led to advanced laminar flow airfoils.

Although NACA deserves credit for its eventual breakthrough in laminar flow wings, the resolution of the issue illustrates a fascinating degree of universality in aeronautical research. NACA—born in response to European progress in aeronautics—benefited through the employment of Europeans like Munk and Theodorsen and profited from a continuous interaction with the European community.

In 1935 Jacobs went to Rome as the NACA representative to the Fifth Volta Congress on High-Speed Aeronautics. During the trip he visited several European research facilities, comparing equipment and discussing the newest theoretical concepts. He concluded that the United States held a leading position, but he asserted that "we certainly cannot keep it long if we rest on our laurels." On his way home, Jacobs stopped off at Cambridge University in Great Britain for long visits with colleagues who were investigating the peculiarities of high-speed flow, including statistical theories of turbulence. These informal exchanges influenced Jacobs' approach to the theory of laminar flow by focusing on the issue of pressure distribution over the airfoil. Working out the details of the idea took three years and engaged the energies of many individuals, including several on Theodorsen's staff, even though he remained skeptical.

Once the theory appeared sound, Jacobs had a wind tunnel model of the wing rushed through the Langley shop and tested it in a new icing tunnel that could be used for some low-turbulence testing. The new airfoil showed a 50 percent decrease in drag. Jacobs was elated, not only because the project incorporated complex theoretical analysis, but also because the subsequent empirical tests justified a new variable density tunnel.

The laminar flow airfoil was used during World War II in the design of the wings for the North American P-51 Mustang, as well as some other aircraft. In operation, the wing did not enhance performance as dramatically as the tunnel tests had suggested. For the best performance, the manufacturing tolerances had to be perfect and the maintenance of wing surfaces needed to be thorough. The rush of mass production during the war and the tasks of meticulous maintenance in combat zones never met the standards of NACA laboratories. Nonetheless, the work on the laminar flow wing pointed the way to a new family of successful high-speed airfoils. These and other NACA wing sections became the patterns for aircraft around the world.

NACA reports began to emerge from an impressive variety of tunnels that went into operation during the 1930s. The refrigerated wind tunnel, declared operational in 1928, became a major tool for the study of ice formation on wings and propellers. In flight, icing exposed a menace to be prevented at all costs. Langley's research in the refrigerated tunnel contributed to successful deicing equipment that not only enabled airliners to keep better schedules in the 1930s but also enabled World War II combat planes to survive many encounters with bad weather. Another facility at Langley, a free-spin wind tunnel, yielded vital information on the spin characteristics of many aircraft; such data were used to improve the maneuverability of aircraft while avoiding deadly spin tendencies. A hydrodynamics test tank solved many riddles for designers of seaplanes and amphibians, by towing hull models to simulated takeoff speeds.

NACA also took a bold look ahead to much higher airplane speeds. In the mid-1930s, when speeds of 200 MPH were quite respectable, the agency proposed a "full-speed" tunnel, which would make it possible to perform tests at a simulated 500 MPH. With an eight-foot diameter, the tunnel allowed tests of comparatively large models, as well as some full-scale components. Completed early in 1936, the eight-foot tunnel played a major role in high-speed aerodynamic research, laying the foundations for later work in high subsonic speeds and the baffling transonic region.

As the research capabilities of NACA expanded, so did the persistent, nagging problems that followed the introduction of successive generations of aircraft. For NACA, one of the most unusual apparitions to appear in the 1930s was the autogiro. First developed by a Spaniard, Juan de la Cierva, in the 1920s,

The P-38 was one of the fighters built in the early 1940s that experienced compressibility effects in steep dives. NACA wartime studies helped designers and pilots cope with the phenomenon and led to intensive postwar research.

the autogiro was thought to have great promise in the immediate future. At first glance it looked like a helicopter, with a huge, multibladed rotor situated above the fuselage. Unlike the helicopter, the autogiro had stubby wings and used a nose-mounted engine with a conventional propeller for forward momentum. The main rotor turned when the craft moved forward, so that its long, thin airfoil blades provided lift, with some assistance from the shortened wings. The autogiro could not take off or land vertically, nor could it hover, but its abbreviated landing and takeoff runs were dramatic, and proponents claimed that the aircraft minimized dangerous stalls. Some writers of the era envisioned the autogiro as a replacement for the family sedan. Accordingly, NACA bought a Pitcairn PCA-2 autogiro (designed and manufactured in Pennsylvania by Harold Pitcairn) and began tests in 1931. These trials did not lead to a permanent niche in American life for the autogiro, but Langley was launched into continuing work on rotary-wing aircraft. Some of the maneuverability tests and other investigations on the autogiro led to testing criteria used even into the 1980s.

The North American XP-51 Mustang was the first aircraft to incorporate an NACA laminar flow airfoil. This is the second XP-51, which arrived at Langley in March 1943.

Flight research like that involving the autogiro became an increasingly valued component of Langley's procedures. Accomplished on an ad hoc basis most of the time, flight testing became more formalized in 1932, when a flight test laboratory appeared at Langley. With separate space allocated for staff, shop work, and an aircraft hangar, the new laboratory made its own contributions to aviation progress during the 1930s.

Among the various airplanes that passed through Langley were two of the most advanced airliners of the era: the Boeing 247 and the Douglas DC-1, which led to the classic DC-3. The Boeing and Douglas designs incorporated the latest aviation technology that had evolved since the end of World War I. With the Ford Tri-Motor of the 1920s, wooden frame and fabric covering had given way to all-metal construction. Unlike the Ford, the Boeing and Douglas transports were low-winged planes with retractable landing gear, and their more powerful twin engines were cowled and mounted into the leading edge of the wings. At 170–80 MPH, they were considerably faster than any of their counterparts, and attention to details like soundproofing and other passenger comforts made them far more popular with travelers. Later versions of the Douglas transport, like the DC-3, added refinements such as wing flaps and variable pitch propellers that made them even more effective in takeoffs and landings and allowed cruising at optimum efficiency at higher altitudes. But it was not clear what would happen if one of the two engines on the new transports failed. At the request of Douglas Aircraft, Langley evaluated the problems of handling and control a twin-engine transport with one engine out. On the basis of these tests, conducted just six months before the DC-3 made its maiden flight, procedures were developed to allow pilots to stay aloft even when one engine malfunctioned, until an emergency landing could be made.

The design revolution leading to all-metal monoplane transports had a similar impact on military aircraft. During 1935 Boeing began flight tests of its huge, four-engined Model 299, the prototype for the B-17 Flying Fortress of World War II. The big airplane's performance exceeded expectations, owing in no small part to design features pioneered by NACA. The Boeing Company sent a letter of appreciation to NACA for its specific contributions to the design of the plane's flaps, airfoil, and engine cowlings. The letter concluded, "it appears your organization can claim a considerable share in the success of this particular design. And we hope that you will continue to send us your 'hot dope' from time to time. We lean rather heavily on the Committee for help in improving our work."

The ability of NACA to carry out the kind of investigations that proved useful was often the result of continuing contacts with the aviation community. One of the most interesting formats for such ideas was the annual aircraft engineering conference, which began in 1926. Attendees included the movers and shakers from the armed services, the aviation press, government agencies, airlines, and manufacturers. These were busy people, and NACA gave them a carefully orchestrated two-day visit to Langley, with plenty of time for conversation.

More than three hundred people made each annual trip, an invitation-only opportunity during the 1930s. NACA's executive secretary, John Victory, became the principal organizer of the event, which had almost sybaritic overtones in the depression era. After gathering in Washington, the group boarded a chartered steamer for a stately cruise down the Chesapeake Bay to Hampton, Virginia. Once ashore, the travelers partook of a generous southern breakfast at a local resort hotel, then headed for Langley in an impressive motorcade of more than fifty cars. The program included reviews of current projects, small-group tours, lab demonstrations, and technical sessions throughout the day. Conference participants motored back to the hotel for cocktails on the veranda, an elaborate banquet, and an overnight return cruise to Washington. Public relations played an obvious role in such outings, but the conferences offered a useful avenue for maintaining contact, for keeping a finger on the pulse of the aviation community, and for keeping the aviation community abreast of NACA's latest research and facilities.

Although NACA personnel may not have enjoyed luxurious perquisites on a daily basis, the agency continued to be a magnet for many young aeronautical engineers. Langley's impressive facilities in particular were a powerful lure, in addition to the opportunity to work closely with well-known people at the cutting edge of flight. Through the 1930s Langley managed to maintain a degree of informality that provided a unique environment for newly hired personnel. John Becker, who reported for duty in 1936, remembered the crowded lunchroom where he found himself rubbing shoulders with the authors of NACA papers that he had just been studying at college. "These daily lunchroom contacts provided not only an intimate view of a fascinating variety of live career models," he wrote, "but also an unsurpassed source of stimulation, advice, ideas, and amusement." By the end of the lunch hour, the white marble tabletops in the lunchroom were invariably covered by sketches, equations, and other miscellany, erased by hand or by a napkin and drawn over again. Becker lamented the loss of this "great unintentional aid to communication" when Langley's growing staff made it necessary to move to a larger, modern cafeteria with unusable table surfaces.

Much of this growth—and the end of an era for Langley and NACA—occurred during the wartime period. In 1938 the total Langley staff numbered 426. Just seven years later, in 1945, Langley had 3,000 personnel.

Military Research

The prewar research at Langley had a catholic fallout, in that the center's activities were applicable to both civil and military aircraft. The commercial aircraft and the fighting planes of the first fifteen years following World War I were very similar in airspeed, wing loading, and general performance. For example, Langley's work on the cowling for radial engines had the encouragement of both civil and military personnel, and NACA cowling eventually appeared on a remarkable variety of light planes, airliners, bombers, and fighter aircraft. Many other NACA projects on icing, propellers, and so on were equally useful to civil and military designs.

About the mid-1930s the phenomenon of mutual benefits began to change. Commercial airline operators put a premium on safety and operational efficiency. Although military designers did not shun such factors, the importance of speed, maneuverability, and operations to very high altitudes meant that NACA research increasingly proceeded along two separate paths. By 1939

This Pitcairn PAA-1 autogiro was flown at Langley for the NACA investigation of an experimental rotor.

the Annual Manufacturers Conference was phased out and replaced by an "inspection," planned solely for representatives of the armed services and delegates from firms having military contracts.

For most of the time after the mid-1930s benchmark, military R&D took the lead in NACA, and the resulting discoveries were incorporated into civilian airplanes. Moreover, there are indications that the U.S. Navy often fared better than the U.S. Army in reaping benefits from Langley's extensive R&D talents. This situation may have stemmed from Langley's early days, when there was some friction about civilian NACA facilities located at the army's Langley Field. Old hands at NACA felt that certain army people wanted to shift NACA's work to McCook Field in Ohio and to conduct all of its operations under an army umbrella. Under the circumstances, the navy appeared to have smoother relations with NACA. At the same time, the navy had reason to rely heavily on NACA's expertise. During the 1920s and 1930s, that service developed its first aircraft carriers. Concurrently, a rather special breed of aircraft had to be developed to fit the demanding requirements of carrier operations. Landings on carriers were bone-jarring events repeated many times (a carrier landing was wryly described as a "controlled crash"); takeoffs were confined to the limited length of a carrier's flight deck. In the process of beefing up structures, improving wing lift, keeping aircraft weight down, enhancing stability and control, and studying other problems, naval aviation and NACA grew up together. Between 1920 and 1935, the navy submitted twice as many research requests as the army.

There were still some instances in which civilian needs benefited military programs. In 1935 Edward P. Warner, Langley's original chief physicist, was working as a consultant for the Douglas Aircraft Company. Warner had the job of determining stability and control characteristics of the DC-4 four-engine transport. Accepted practice of the day usually meant informal discussions between pilots and engineers as the latter tried to design a plane having the often elusive virtues of "good flying qualities." At Warner's request NACA began a special project to investigate flying qualities desired by pilots so that numeric guidelines could be written into design specifications. At Langley, researchers used a specially instrumented Stinson Reliant to develop usable criteria. Measurable control inputs from the test pilot were correlated with the plane's design characteristics to develop a numeric formula that could be applied to other aircraft. Further tests on twelve different planes gave a comprehensive set of figures for both large and small aircraft. As military programs gained urgency in the late 1930s, the formulas for flying qualities were increasingly used in the design of new combat planes.

The growing international threat found the American aviation industry in far better shape than it was on the eve of World War I. In terms of civil aviation, the United States had established an enviable record of progress. Commercial airliners like the DC-3 had set a world standard and were widely used by many foreign airlines on international routes. Airline operations had reached new levels of maturity, not only in marketing and advertising to attract a growing clientele, but also in a variety of supporting activities. These included maintenance and overhaul procedures, radio communication, weather forecasting, and long-distance flying. Many of these skills proved valuable to the military after the outbreak of war. Pan American World Airways (Pan Am), which had pioneered long-distance American routes throughout the Caribbean, the Pacific, and the Atlantic, shared its skills and personnel to help the Air Transport Command evolve a remarkable global network during the war years. Pan Am relied on a series of impressive flying boats designed and built by Sikorsky, Martin, and Boeing during the 1930s. Although the military airlift services depended more on landplanes like the DC-3 (the military version was known as the C-47) and the DC-4 (or C-54), many of the imaginative design concepts of the flying boats pointed the way to the multi-engine airliners that replaced them.

NACA carried out single-engine performance tests on the Douglas DC-3, as well as studies related to stall characteristics and the effects of icing.

There were even benefits for the light plane industry. Despite the depression, personal and business flying became firmly entrenched in the American aviation scene. Manufacturers offered a surprising array of designs, from the economical two-place Piper Cub J-3 to the swift four- to five-place business planes produced by Stinson and Cessna. At the top of the scale, the Beech D-18, a twin-engine speedster, supplied the era's ultimate in corporate transportation. When war came, these and other manufacturers were ready to turn out the dozens of primary trainers (larger planes for navigational and bombing instruction) and various components that made up the other equipment in the U.S. armed forces.

The air force itself was beginning to receive the type of combat planes that enabled it to meet aggressive fliers in the skies over Europe and the Far East. Prewar fighters like the Curtiss P-40 soon gave way to the Lockheed P-38, the Republic P-47, and the North American P-51. A new family of medium bombers and heavy bombers included the redoubtable B-17 Flying Fortress,

NACA continued drag reduction studies on the navy's Brewster F2A-2 Buffalo. This F2A-2 arrived at Langley from the factory by truck in 1942 and was sent to NAS Norfolk two years later.

derived from the Boeing 299. Aboard the U.S. Navy's big new aircraft carriers, biplanes had been replaced by powerful monoplanes like the Grumman Wildcat, and then the Hellcat and the Vought Corsair. There were also new dive bombers and long-legged patrol planes like the Catalina amphibian. Directly or indirectly, the majority of these aircraft profited from NACA's productivity during the 1930s as well as during the war.

The War Years

Even though Langley and NACA had contributed heavily to the progress of American aviation, there were still some members of Congress who had never heard of them. Before World War II, a series of committee reports brought a dramatic change. During the late 1930s, John Jay Ide, who manned NACA's listening post in Europe, reported unusually strong commitments to aeronautical research in Italy and Germany, where no fewer than five research centers were under development. Germany's largest, located near Berlin, had a reported 2,000 personnel at work, compared to Langley's 350. Although the Fascist powers were developing civilian aircraft, it became apparent that military research absorbed the lion's share of work at the new centers. Under the circumstances, NACA formed stronger alliances with military services in the United States for expansion of its own facilities.

In 1936 the agency put together a special committee on the relationship of NACA to national defense in time of war, chaired by the chief of the Army Air Corps, Major General Oscar Westover. Its report, released two years later, called for expanded facilities in the form of a new laboratory—an action underscored by Charles Lindbergh, who had just returned from a European tour warning that Germany clearly surpassed America in military aviation. A follow-up committee, chaired by Rear Admiral Arthur Cook, chief of the navy's Bureau of Aeronautics, recommended that the new facility be located on the West Coast, where it could work closely with the growing aircraft industry in California and Washington. After congressional debate, NACA received money for expanded facilities at Langley (pacifying the Virginia congressman who ran the House Appropriations Committee) along with a new laboratory at Moffett Field, south of San Francisco. The official authorization came in August 1939; only a few weeks later, German planes, tanks, and troops invaded Poland. World War II had begun.

The outbreak of war in Europe, coupled with additional warnings from NACA committees and from Lindbergh about American preparedness, triggered support for a third research center. British, French, and German military planes were reportedly faster and more able in combat than their American counterparts. Part of the reason, according to experts, was the European emphasis on liquid-cooled engines that yielded benefits in speed and high altitude operations. In the United States, the country's large size had influenced the development of air-cooled engines that were more suited to longer ranges and fuel efficiency. Moreover, according to Lindbergh, NACA's earlier agreement to leave engine development to the manufacturers left the country with inadequate national research facilities for aircraft engines. Congress quickly responded, and an Aircraft Engine Research Laboratory was set up near the municipal airport in Cleveland, Ohio. This third new facility in the Midwest gave NACA a geographical balance, and the location also put it in a region that already had significant ties to the power-plant industry.

The site at Moffett Field became Ames Aeronautical Laboratory in 1940, in honor of Dr. Joseph Ames, charter member of NACA and its long-time chairman. The "Cleveland laboratory" remained just that until 1948, when it was renamed the Lewis Flight Propulsion Laboratory, in memory of its veteran director of research, George Lewis. Key personnel for both new laboratories came

Many women joined NACA during World War II. Here technicians prepare a wind tunnel model, a one-twelfth-scale model of a flying boat wing, for realistic tests.

from Langley, and the two junior labs tended to defer to Langley for some time. By 1945, after several years of managing their own wartime projects, the Ames and Cleveland laboratories felt less like adolescents and more like peers of Langley. NACA, like NASA after it, became a family of labs, but with strong individual rivalries.

In the meantime, the requirements of national security took priority. One significant project undertaken on the eve of World War II demonstrated the work at Langley that had a major influence on aircraft design for years afterward. During 1938 the navy became frustrated with the performance of a new fighter, the XF2A Brewster Buffalo. After the navy flew a plane to Langley, technicians set it up in the full-scale tunnel for drag tests. It took only five days to uncover a series of small negative aspects of the plane's design.

To the casual eye, the 250-MPH fighter with retractable gear appeared aerodynamically "clean." But the wind tunnel evaluations pinpointed many specific design features that created drag. The exhaust ports, the gunsight, the guns, and the landing gear all protruded into the slipstream during flight; the accumulated drag effects hampered the plane's performance. By revamping these and other areas, NACA reported a 10 percent increase in speed. Such a performance improvement, without raising engine power or reducing fuel efficiency, immediately caught the attention of other designers. Within the next two years, no fewer than eighteen military prototypes went through the "clean-up" treatment given to the XF2A. Even though the Brewster Buffalo failed to win an outstanding combat record, others did, including the Grumman XF4F Wildcat, the Republic XP-47 Thunderbolt, and the Chance Vought XF4U Corsair. The enhanced performance of these planes often made the difference between victory and defeat in air combat. Moreover, specialists in the analysis of engine cooling and duct design later set the guidelines for inducing air into a postwar generation of jet engines.

The pace of war created personnel problems, especially when Selective Service began to claim qualified males after 1938. In the early years of the war, NACA personnel officers did considerable traveling each month to get deferments for employees working on national defense projects. Nonetheless, NACA sometimes lost more employees than it was able to recruit. The issue was not resolved until early in 1944, when all eligible Langley employees were inducted into the Air Corps Enlisted Reserves, then put on inactive status under the exclusive management of NACA. NACA draftees were given honorable discharges after Japan's surrender in 1945. The issue of the draft was not a threat to women, who made up about one-third of the entire staff by the end of the war. Although most of the female employees held traditional jobs as secretaries, increasing numbers had technical positions in the laboratories. Some did drafting and technical illustrating; some did strain-gauge measurements; others made up entire computing groups who worked through the reams of figures pouring out of the various wind tunnels. A few held engineering posts. Women at Langley may not have advanced as rapidly in civil service as their male counterparts, but most of the female employees later recalled that they were treated better at NACA than other contemporary employers generally treated women employees.

Over the course of the war years, NACA's relationship with industry underwent a fundamental change. Since its inception the agency had refused to have an industry representative sit on the main committee, fearing that industry influence would make NACA into a "consulting service." But the need to respond to industry goals in the emergency atmosphere of war led to a change in policy. The shift came in 1939, when George Mead became vice chairman of NACA and chairman of the Power Plants Committee. Mead had recently retired as a vice president of the United Aircraft Corporation, and his position in NACA, considering his high-level corporate connections, reflected a new trend. During the war dozens of corporate representatives descended on Langley to observe and actually assist in testing. In the process they forged additional direct links between NACA and aeronautical industries.

Much of the wartime work involved refinement of manufacturers' designs, ranging from fighters through bombers like the B-29. Aircraft as large as the B-29 design were not tested as full-sized planes, but considerable data was generated from models. During 1942 the B-29 design was thoroughly investigated in Langley's eight-foot high-speed tunnel, and Boeing engineers heaped praise on Langley technicians for their cooperation and the high quality of the data generated by the tests.

Despite the success of American warplanes, two of the major aeronautical trends of the era nearly escaped NACA's attention. The agency endured much criticism in the postwar era for its apparent lapses in the development of jet propulsion and in the area of high-speed research leading to swept wings. America's rapid postwar progress in these fields suggests that there may have been a lapse of sorts, although not as total as many critics believed.

Rocketry

Nothing in the original NACA charter charged it with responsibility for research in rocketry. Some of NACA's personnel had a personal interest in rocket technology, but most early developments in this field came from sophisticated amateur associations, such as the American Interplanetary Society. During World War II, governments suddenly became more interested in rocketry as a powerful new weapon.

The existence of organized groups like the VfR in Germany signaled the increasing fascination with modern rocketry in the 1930s, and information was exchanged frequently between the VfR and groups such as the British Interplanetary Society (1933) and the American Interplanetary Society (1930). Even Goddard occasionally had correspondence printed in the American Interplanetary Society's *Bulletin*, but he remained aloof from other American researchers, cautious about his results, and concerned about patent infringements. Because of Goddard's reticence, in contrast to the more visible personalities in the VfR, and because of the publicity given the German V-2 of World War II, the work of British, American, and other groups during the 1930s has been overshadowed. Their work, if not as spectacular as the V-2 project, nevertheless contributed to the growth of rocket technology in the prewar era and to the successful use of a variety of Allied rocket weapons in World

War II. Although organizations like the American Interplanetary Society (which became the the American Rocket Society in 1934) succeeded in building and launching several small chemical rockets, much of the significance of such societies lay in their role as the source of a growing number of technical papers on rocket technologies.

But rocket development was complex and expensive. The cost and the difficulties of planning and organization meant that, sooner or later, the major work in rocket development would have to occur under the aegis of permanent government agencies and government-funded research bodies. In America, significant team research began in 1936 at the Guggenheim Aeronautical Laboratory, California Institute of Technology, or GALCIT. In 1939 this group received the first federal funding for rocket research and had special success with rockets to assist aircraft takeoff. The project was known as JATO, for jet-assisted takeoff, since the word *rocket* still carried negative overtones in many bureaucratic circles. JATO research led to substantial progress in a variety of rocket techniques, including both liquid and solid propellants. Work in solid propellants proved especially fortuitous for the United States; during World War II American armed forces made wide use of the bazooka (an antitank rocket) as well as barrage rockets (launched from ground batteries or from ships) and high-velocity air-to-surface missiles.

The most striking achievements in rocketry originated in Germany. In the early 1930s the VfR attracted the attention of the German army, since armament restrictions imposed by the Treaty of Versailles at the conclusion of World War I left the door open for rocket development. A military team began rocket research as a variation of long-range artillery. One of the chief assistants was an enthusiast from the VfR, Wernher von Braun, now twenty-two years old, who joined the army organization in October 1932. By December the army rocket group had static-fired a liquid propellant rocket engine at the army's proving grounds near Kummersdorf, south of Berlin. During the next year it became evident that the test and research facilities at Kummersdorf would not be adequate for the scale of the hardware under development. A new location, shared jointly by the German army and air force, was developed at Peenemünde, a coastal area on the Baltic Sea. Starting with 80 researchers in 1936, there were nearly 5,000 personnel at work by the time of the first launch of the awesome, long-range V-2 in 1942. Later in the war, with production in full swing, the workforce swelled to about 18,000.

After completing his doctorate in 1934 (on rocket combustion), von Braun became the leader of a formidable R&D team in rocket technology at Peenemünde. Like so many of his cohorts in original VfR projects, von Braun still harbored an intense interest in rocket development for manned space travel. Early in the V-2 development agenda, he began looking at the rocket in terms of its promise for space research as well as its military role, but he found it prudent to adhere rigidly to the latter. Paradoxically, German success in the wartime V-2 program became a crucial legacy for postwar American space efforts.

Chapter 3

Jets, Sonic Speed, and Satellites, 1945–1958

At the end of World War II, the combined aviation elements of the U.S. Army Air Force (and the atomic bomb), the U.S. Marines, and the U.S. Navy made America's air forces the most formidable in the world—but not necessarily the most advanced. Early American military jets relied on gas turbine engines developed by the British, and significant progress in swept-wing designs owed a tangible debt to German wartime research. Spurred by cold war concerns, the impact of well-funded flight research and intensive scientific efforts buttressed an ambitious agenda within NACA. The first supersonic aircraft rested primarily on American rocket research and on indigenous aerodynamic investigations. Nonetheless, after the Soviet Union successfully orbited the world's first artificial satellite in 1957, critics assailed NACA's apparent failure to stay on the cutting edge of space technology. Consequently NACA experienced a complete reorganization and a new emphasis on astronautics, in order to achieve leadership in the new arena of space exploration and cold war competition. As a born-again agency, the National Aeronautics and Space Administration began operations in 1958.

Into the Jet Age

On 1 October 1942, the Bell XP-59A, America's first jet plane, took to the air over a remote area of the California desert. There were no official NACA representatives present. NACA did not even know the aircraft existed, and the engine was based entirely on a top-secret British design. After the war, the failure of the United States to develop jet engines, swept-wing aircraft, and supersonic designs was generally blamed on NACA. Critics argued that NACA, as America's premier aeronautical establishment (the one that presumably led the world in successful aviation technology), had somehow allowed leadership to slip to the British and the Germans during the late 1930s and during World War II.

In retrospect, the NACA record seems mixed. There were some areas, such as gas turbine technology, in which the United States clearly lagged, although NACA researchers had begun to investigate jet propulsion concepts. There were other areas, including swept-wing designs and supersonic aircraft, in which NACA had made important forward steps. Unfortunately, the lack of advanced propulsion systems, such as jet engines, made such investigations academic exercises. NACA's forward steps undeniably trailed the rapid strides made in Europe.

During the 1930s aircraft speeds of 300–350 MPH represented the norm, and designers were already thinking about planes able to fly at 400–450 MPH. At such speeds the prospect of gas turbine propulsion became compelling. With a piston engine, the efficiency of the propeller began to fall off at high speeds, and the propeller itself produced a significant drag factor. The problem was to obtain sufficient R&D funds for what seemed to be unusually exotic gas turbine power plants.

In England, Royal Air Force (RAF) officer Frank Whittle doggedly pursued research on gas turbines through the 1930s, eventually acquiring some funding through a private investment banking firm after the British Air Ministry turned him down. Strong government support finally materialized on the eve of World War II, and the single-engine Gloster experimental jet fighter flew in the spring of 1941. English designers leaned toward the centrifugal-flow jet engine, a comparatively uncomplicated gas turbine design, and a pair of these power plants equipped the Gloster Meteor of 1944. Although Meteors entered RAF squadrons before the end of the war and shot down German V-1 flying bombs, the only jet fighter to fly in air-to-air combat came from Germany—the Me-262. Hans von Ohain, a researcher in applied physics and aerodynamics at the University of Göttingen, had unknowingly followed a course of investigation that paralleled Whittle's work; he took out a German patent on a centrifugal engine in 1934. Research on gas turbine engines evolved from several other sources shortly thereafter, and the German Air Ministry, using funds from Hitler's rearmament program, earmarked more money for this research. Although a centrifugal type powered the world's first gas turbine aircraft flight by

The North American F-86 featured swept wings and tail surfaces. This F-86 aircraft is mounted in the 40-by-80-foot full-scale wind tunnel at NACA Ames Aeronautical Laboratory, Moffett Field, California.

the He-178 in 1939, the axial-flow jet, more efficient and capable of greater thrust, was used in the Me-262 fighters that entered service in the autumn of 1944.

In the United States, the idea of jet propulsion had surfaced as early as 1923, when an engineer at the Bureau of Standards wrote a paper on the subject, which was published by NACA. The paper came to a negative conclusion: fuel consumption would be excessive, compressor machinery would be too heavy, and high temperatures and high pressures were major barriers. These were assumptions that subsequent studies and preliminary investigations seemed to substantiate into the 1930s. By the late 1930s, the Langley staff became interested in the idea of a form of jet propulsion to augment power for military planes for takeoff and during combat. In 1940 Eastman Jacobs and a small staff came up with a jet propulsion test bed they called the Jeep. This was a ducted-fan system that used a piston engine power plant to combine the engine's heat and exhaust with added fuel injection for brief periods of added thrust, much like an afterburner. A test rig was in operation during the spring of 1942. By the summer, however, the Jeep had grown into something else—a research aircraft for transonic flight. With Eastman Jacobs again, a small team made design studies of a jet plane having the ducted-fan system completely closed within the fuselage, similar to the Italian Caproni-Campini plane that flew in 1942. Although work on the Jeep and the jet plane design continued into 1943, these projects had already been overtaken by European developments.

During a tour to Britain in April 1941, General H. H. "Hap" Arnold, chief of the U.S. Army Air Force, was dumbfounded to learn about a British turbojet plane, the Gloster E28/39. The aircraft had already entered its final test phase and, in fact, made its first flight the following month. Fearing a German invasion, the British were willing to share the turbojet technology with America. That September an air force major, with a set of drawings manacled to his wrist, flew from London to Massachusetts, where General Electric went to work on an American copy of Whittle's turbojet. An engine, along with Whittle himself, followed. The development of the engine and the design of the Bell XP-59 were so cloaked in secrecy that NACA learned nothing about them until the summer of 1943. Moreover, the design of the Lockheed XP-80, America's first operational jet fighter, was already under way.

General Arnold may have lost confidence in NACA's potential for advanced research when he stumbled onto the British turbojet plane. It may be that British and American security requirements were so strict that the danger of sharing information with the civilian agency, where the risk of leaks was magnified, justified Arnold's decision to exclude NACA. The answers were not clear. In any case, the significance of turbojet propulsion and rising speeds magnified the challenges of transonic aerodynamics. This was an area where NACA had been at work for some years, though not without influence from overseas.

Shaping New Wings

As information on advanced aerodynamics began to trickle out of defeated Germany, American engineers were impressed. Photographs of some of the startling German aircraft, like the batlike Me-163 rocket powered interceptor and the improbable Junkers JU-287 jet bomber, with its forward-swept wings, prompted critics to ask why American designs appeared to lag behind the Germans'. It seemed to be the story of the turbojet again. The vaunted NACA had let advanced American flight research fall precariously behind during the war. True, the effect of wartime German research made an impact on postwar American development of swept wings, leading to high performance jet bombers like the Boeing B-47 and the North American F-86 jet fighter. But American engineers, including NACA personnel, had already made independent progress along the same design path when the German hardware and drawings were turned up at the end of World War II.

Like several other chapters in the story of high-speed flight, the story began in Europe, where an international conference on this topic—the Volta Congress—met in Rome during October 1935. Among the participants was Adolf Busemann, a young German engineer from Lübeck. As a youngster he had watched innumerable ships navigating Lübeck's harbor, each vessel moving within the V-shaped wake trailing back from the bow. When he became an aeronautical engineer, this image was a factor that led him to consider designing an airplane with swept wings. At supersonic speeds, the wings would function effectively inside the shock waves stretching back from the nose of an airplane. In the paper Busemann presented at the Rome conference, he analyzed this phenomenon and predicted that his "arrow wing" would have less drag than straight wings exposed to the shock waves.

There was polite discussion of Busemann's paper but little else, since the propeller-driven aircraft of the 1930s lacked the performance to merit serious consideration of such a radical design. Within a decade the emergence of the turbojet dramatically changed the picture. In 1942 designers for the Messerschmitt firm, builders of the remarkable Me-262 jet fighter, saw the potential of swept-wing aircraft and studied Busemann's paper more intently. After promising wind tunnel tests, Messerschmitt had a swept-wing research plane under development, but the war ended before the plane was finished.

In the United States, progress toward swept-wing design proceeded independently of the Germans, although admittedly behind them. The American chapter of the swept-wing story began with Michael Gluhareff, a graduate of the Imperial Military Engineering College in Russia during World War I. He fled the Russian revolution and gained aeronautical engineering experience in Scandinavia. Gluhareff arrived in the United States in 1924 and joined the company of a Russian compatriot, Igor Sikorsky. By 1935 Gluhareff was chief of design for Sikorsky Aircraft and eventually became a major figure in developing the first practical helicopter. He became fascinated by the possibilities of low-aspect ratio tailless aircraft and built a series of flying models in the late 1930s. In a memo to Sikorsky in 1941, he described a possible pursuit-interceptor with a delta-shaped wing swept back at an angle of 56 degrees. Gluhareff reasoned that the delta wing would delay the phenomenon known as the compressibility effect, creating drag, and make the delta wing adaptable to exceptionally high speeds.

Eventually, a wind tunnel model made its debut; initial tests were encouraging. But the army declined to follow up because there were several other unconventional projects already under way. Fortunately, a business associate of Gluhareff kept the concept alive by using the Dart design, as it was called, as the basis for an air-to-ground glide bomb in 1944. This time, the army was intrigued and asked NACA to evaluate the project. Thus, a balsa model of the Dart, along with some data, wound up on the desk of Robert T. Jones, a Langley aerodynamicist.

Jones had a reputation as a bit of a maverick. A college dropout, he signed on as a mechanic for a barnstorming outfit known as the Marie Meyer Flying Circus. Jones became a self-taught aerodynamicist who could not find a job during the 1930s depression. He moved to Washington, D.C., and worked as an elevator operator in the Capitol. There he met a congressman who paid Jones to tutor him in physics and mathematics. Impressed by Jones' abilities, the legislator got him into a Works Projects Administration program that led to a job at Langley in 1934. With his innate intelligence and impressive intuitive abilities, Jones quickly moved ahead in NACA hierarchy.

Studying Gluhareff's model, Jones soon realized that the lift and drag figures for the Dart were based on outmoded calculations for wings of high-aspect ratio. Using more recent theory for low-aspect ratio shapes, backed by some theoretical work done by Max Munk, Jones predicted that wings with subsonic sweep could effectively fly at speeds greater than Mach 1 (that term, meaning the speed of sound, comes from the name of the Austrian physicist Ernst Mach). He reported the new theory to NACA's research director, George Lewis, early in 1945.

So much for theory. Only testing would provide the data to make or break Jones' concept. Langley personnel went to work, fabricating two small models

This group portrait displays typical high-speed research aircraft that made headlines at Muroc in the 1950s. The photo shows the X-3 (center) and, clockwise from left: X-1A, the third D-558-1. XF-92A, X-5, D-558-2, and X-4.

to see what would happen. Technicians mounted the first model on the wing of a P-51 Mustang. The plane's pilot took off and climbed to a safe altitude before nosing over into a high-speed dive toward the ground. In this attitude, the accelerated flow of air over the Mustang's wing went supersonic, and the instrumented model on the plane's wing began to generate useful data. For wind tunnel tests, the second model emerged as a truly diminutive article, crafted of sheet steel by Jones and two other engineers. Langley's supersonic tunnel had a 9-inch throat, so the model had a 1.5-inch wingspan, in the shape of a delta. The promising test results, issued 11 May 1945, were released before Allied investigators in Europe had the opportunity to interview German aerodynamicists on delta shapes and swept-wing developments.

Jones was already at work on variations of the delta, including his own version of the swept-wing configuration. Late in June 1945, he published a summary of this work as NACA Technical Note Number 1033. Jones suggested that the proposed supersonic plane under development should have swept wings, but designers opted for a more conservative approach. Other design staffs were fascinated by the promise of swept wings, especially after the appearance of the German aerodynamicists in America.

The Germans arrived courtesy of Operation Paperclip, a high-level government plan to scoop up leading German scientists and engineers during the closing months of World War II. Adolf Busemann eventually wound up at NACA's Langley laboratory, and scores of others joined air force, army, and contractor staffs throughout the United States. Information from the research done by Robert Jones had begun to filter through the country's aeronautical community before the Germans arrived. Their presence, along with the obvious progress seen in advanced German aircraft produced by 1945, bestowed the imprimatur of proof to swept-wing configurations. At Boeing, designers at work on a new jet bomber tore up the sketches for a conventional plane with straight wings and built the B-47 instead. With its long, swept wings, the B-47 launched Boeing into the development of a remarkably successful family of swept-wing bombers and jet airliners. At North American, a conventional jet fighter with straight wings, the XP-46, went through a dramatic metamorphosis, eventually taking to the air as the famed F-86 Sabre, a swept-wing fighter that racked up an enviable combat record during the Korean conflict in the 1950s.

Nonetheless, America had been demonstrably lagging in jets and swept-wing aircraft in 1945, and NACA became the target of criticism from postwar congressional and air force committees. NACA may not have been as bold as it could have been, or the agency may have been so caught up in immediate wartime improvements that crucial areas of basic research received short shrift. There were administrative changes in response to these issues. In any case, as historian Alex Roland noted in his study of NACA, its shortcomings "should not be allowed to mask its real and significant contributions to American aerial victory in World War II." Moreover, NACA's postwar achievements in supersonic research and rapid transition into astronautics reflected a new vigor and momentum.

Going Supersonic

During World War II the increasing speeds of fighter aircraft began to create new problems. The Lockheed P-38 Lightning, for example, could exceed 500 MPH in a dive. In 1941 a Lockheed test pilot died when shock waves from the plane's wings (where the airflow over the wings reached 700 MPH) created turbulence that tore away the horizontal stabilizer, sending the plane into a fatal plunge. From wind tunnel tests, researchers knew something about the shock waves occurring at Mach 1, the speed of sound. The phenomenon was obviously attended by danger. Pilots and aerodynamicists alike muttered about the threatening dimensions of what came to be called the sound barrier.

Researchers faced a dilemma. In wind tunnels, with models exposed to near-sonic velocities, shock waves began bouncing from the tunnel walls, the "choking" phenomenon, resulting in questionable data. In the meantime high-speed combat maneuvers brought additional reports of loss of control caused by turbulence and, in several cases, crashes involving planes whose tails had wrenched loose in a dive. Since data from wind tunnels remained unreliable, researchers proposed a new breed of research plane to probe the sound barrier. Two of the leaders were Ezra Kotcher, a civilian on the air force payroll, and John Stack, on NACA staff at Langley.

By 1944 John Stack and his NACA research team proposed a jet powered aircraft, a conservative, safe approach to high-speed flight tests. Kotcher's group wanted a rocket engine, a more dangerous alternative, with explosive fuels aboard, but more likely to achieve the high velocity needed to reach the speed of sound. The air force had the funds, so Stack and his colleagues agreed. The next problem involved design and construction of the rocket plane.

The contract eventually went to Bell Aircraft Corporation in Buffalo, New York. The company had a reputation for unusual designs, including the first American jet, the XP-59A Airacomet. The designer was Robert J. Woods, who had worked with John Stack at Langley in the 1920s before he joined Bell Aircraft. Woods had close contacts with NACA as well as the air force. During a casual visit to Kotcher's office at Wright Field, Woods agreed to design a research plane capable of reaching 800 MPH at an altitude of thirty-five thousand feet. Woods then called his boss, Lawrence Bell, to break the news. "What have you done?" Bell lamented, only half in jest.

The Bell design team worked closely with the air force and NACA. This was the first time that the Langley staff had been involved in the initial design and construction of a complex research plane. Even with the air force bearing the cost and sharing the research load, this collaboration marked a significant departure from former NACA procedures. For the most part, design issues were amicably resolved, although some questions caused heated exchanges. The wing design created one such controversy, until Langley thoroughly analyzed the issue and advised that a thin cross section would be the most promising to achieve supersonic speed. As the design of the plane progressed, Bell's engineers came up with a plane that measured only 31 feet long with a wingspan of just 28 feet.

Research at Langley influenced other aspects of the design. Realizing that turbulence from the wing might create control problems around the tail, John Stack advised Bell to place the horizontal stabilizer on the fin, above the turbulent flow. He also recommended a stabilizer that was thinner than the wing, ensuring that shock waves would not form on the wing and the tail at the same time and thereby improving the pilot's control over the accelerating aircraft. In making these decisions, the design team recognized that not much was known about the flight speeds for which the plane was intended. There was some interesting aerodynamic information available on the .50 caliber bullet, however, so the fuselage shape became keyed to ballistics data from this unlikely source. With this image in mind, designers placed the cockpit under a canopy that matched the rounded contours of the fuselage, since a conventional design atop the fuselage would create too much drag.

The engine constituted one of the few really exotic aspects of the supersonic plane. The jet engines under development fell far short of the thrust required to reach Mach 1, forcing designers to consider rocket engines, a radical new technology at that time. The original engine candidate came from a small Northrop design for a flying wing. The propellants, red fuming nitric acid and aniline, ignited spontaneously when mixed. Curious about this volatile combination, some Bell engineers obtained some samples, put the stuff in a pair of bottles taped together, found some isolated rocks outside the plant, and tossed the bottles into them. The fierce eruption that followed left them aghast. Considering the consequences to the plane and its pilot in case of a landing accident or a fuel leak, they decided that a different propulsion system would have to be used. They settled on a rocket engine supplied by a pioneering outfit aptly named Reaction Motors, Inc. The engine burned a mixture of alcohol and distilled water along with liquid oxygen to produce a thrust of fifteen hundred pounds from each of four thrust chambers. Because the propellant capacity of the research plane was limited, the design team decided to use a Boeing B-29 Superfortress to carry it to about twenty-five thousand feet. After dropping from the B-29 bomb bay, the pilot would ignite the rocket engine for a high-speed dash; when all its fuel was consumed, the plane would have to glide earthward and make a dead-stick landing. By this time the plane had acquired the designation XS-1, for Experimental Sonic 1, soon shortened to X-1 by those associated with it.

By autumn 1946 the X-1 managers organized a clandestine operation for shipping it to a remote air base in California's Mojave Desert—Muroc Army Air Field, familiarly known as Muroc, after a small settlement on the edge of Rogers Dry Lake. This was the air force flight test center, an area of three hundred square miles of desolation in the California desert northwest of Los An-

This Grumman F-11, developed for the navy in the 1950s, was one of the first aircraft to incorporate the area-ruled fuselage into its design.

geles. Muroc, which had begun as an air force bombing and gunnery range, was a suitably secure and remote location. Moreover, the concrete-hard lake bed was well suited for experimental testing. Test aircraft sometimes made emergency landings, and the barren miles of Rogers Dry Lake allowed these unscheduled approaches from almost any direction. This austere, almost surrealistic desert setting provided an appropriate environment for a growing roster of exotic planes based there in the postwar years.

The X-1 arrived under a cloud of gloom from overseas. The British had also been developing a plane to pierce the sound barrier, the de Havilland D.H. 108 Swallow, a swept-wing, jet propelled, tailless airplane. Geoffrey (son of the firm's founder, Geoffrey de Havilland) died during a high-speed test of the sleek aircraft in September 1946. The barrier represented a deadly threat.

Through the end of 1946 and into the autumn of 1947, the test flights took the X-1 to higher and higher speeds, past Mach .85, the region where statistics on subsonic flight more or less faded away. On the one hand, the X-1 test crew felt increasing confidence that their plane could successfully make the historic run. On the other hand, NACA engineers, such as Walt Williams, grudgingly admitted to a very unsettled feeling as the store of known data began to run out.

The air force and NACA put considerable trust in the piloting skills of Captain Charles "Chuck" Yeager, a World War II fighter ace. During the test sequences, he learned to keep his exuberance under control and to acquire a thorough knowledge of the X-1's quirks. On the morning of 14 October 1947, the day of the supersonic dash, Yeager's aggressive spirit helped him overcome the discomfort of two broken ribs from a horseback accident a few days earlier. A close friend helped the wincing Yeager into the cramped cockpit, then slipped him a length of broom handle so that he could secure the safety latch. Yeager reported that he was ready to go. At 20,000 feet above the desert, the X-1 dropped away from the B-29.

Yeager fired up the four rocket chambers and shot upward to 42,000 feet. Leveling off, he shut down two of the chambers while making a final check of the plane's readiness. Already flying at high speed, Yeager fired a third chamber and watched the instruments jump as buffeting occurred. Then the flight smoothed out; needles danced ahead as the X-1 went supersonic. Far below, test personnel heard a loud sonic boom slap across the desert. The large data gap bothering Walt Williams had just been filled in.

A need for high-speed wind tunnel tests still existed. In the seven-by-ten-foot tunnel at Langley, technicians built a hump in the test section; as the airstream accelerated over the hump, models could be tested at Mach 1.2 before the "choking" phenomenon occurred. A research program came up with the idea of absorbing the shock waves by means of longitudinal openings, or slots, in the test section. The slotted-throat tunnel became a milestone in wind tunnel evolution, permitting a full spectrum of transonic flow studies. In another high-speed test program, Langley used rocket propelled models, launching them from a new test facility at Wallops Island, north of Langley on the Virginia coast. This became the Pilotless Aircraft Research Division (PARD), established in the autumn of 1945. During the next few years, PARD used rocket boosters to make high-speed tests on a variety of models of new planes under development. These included most of the subsonic and supersonic aircraft flown by the armed services during the decades after World War II. In the 1960s PARD facilities supported the Mercury, Gemini, and Apollo programs as well.

As full-sized aircraft took to the air, new problems inevitably cropped up. Researchers soon realized that a sharp increase in drag occurred in the transonic region. Slow acceleration through this phase of flight consumed precious fuel and also created control problems. At Langley, Richard T. Whitcomb became immersed in the problem of transonic drag. In the course of his analysis, Whitcomb developed a hunch that the section of an airplane where the fuselage joined the wing was a key to the issue. After listening to some comments by Adolph Busemann on airflow characteristics in the transonic regime, Whitcomb hit upon the answer to the drag problem—the concept of the area rule.

Essentially, the area rule postulated that the cross section of an airplane should remain reasonably constant from nose to tail, minimizing disturbance of the airflow and drag. But the juncture of the wing root to the fuselage of a typical plane represented a sudden increase in the cross-sectional area, creating the drag that produced the problems encountered in transonic flight. Whitcomb's solution was to compensate for this added wing area by reducing the area of the fuselage. The result was the "wasp-waisted" look, often called the "Coke bottle" fuselage. Almost immediately, it proved its value. A new fighter, Convair's XF-102, was designed as a supersonic combat plane but repeatedly frustrated the efforts of test pilots and aerodynamicists to achieve its design speed. Rebuilt with an area rule fuselage, the XF-102 sped through the transonic region like a champion; the Coke-bottle fuselage became a feature on many high performance aircraft of the era: the F-106 Delta Dart (successor to the F-102), the Grumman F-11, the Convair B-58 Hustler bomber, and others.

For flight testing the "NACA Muroc Flight Test Unit" usually took advantage of the hardpan surrounding the Muroc Air Force Base in California. Several name changes took place over the years, including the rechristening of Muroc as Edwards Air Force Base in 1950. Changes in NACA and NASA designations also occurred, leading to the establishment of NASA's Flight Research Center at the facility in 1959. Through it all, personnel from the air force as well as NACA/NASA often referred to the site as Muroc and, later, simply as Edwards.

A succession of X aircraft, designed primarily for flight experiments, populated the skies above Muroc in a continuous cycle of R&D. Two more X-1 aircraft were ordered by the air force, followed by the X-1A and the X-1B, which investigated thermal problems at high speeds. The navy used the Muroc flight test area for the subsonic jet powered Douglas Skystreak, accumulating air-load measurements unobtainable in early postwar wind tunnels. The Skystreak was followed by the Douglas Skyrocket, a swept-wing research jet. (Later it was equipped with a rocket engine that surpassed twice the speed of sound for the first time in 1953.) The Douglas X-3, which fell short of expectations for further flight research in the Mach 2 range, nevertheless yielded important design insights on the phenomenon of inertial coupling (solving a control problem for the North American F-100 Super Sabre) and on the structural use of titanium (incorporated in the X-15 and other subsequent supersonic fighter designs) and data that were applied in the design of the Lockheed F-104 Starfighter. NACA kept involved throughout these programs. In a number of ways, the X aircraft contributed substantially to the solution of high-speed flight conundrums and enhanced the design of future jet airliners, establishing a record of consistent progress aside from the speed records that so fascinated the public.

Although much of NACA's work in this era had to do with military aviation, a good number of aerodynamic lessons were applicable to nonmilitary research planes and to civilian aircraft. This was true of the XB-70 supersonic bomber, which yielded useful data for jet transports that operated in the transonic region. Helicopters, introduced into limited combat service at the end of World War II, entered both military and civilian service in the postwar era. NACA flight-tested new designs to help define handling qualities. Using wind tunnel experience, researchers also developed a series of special helicopter airfoil sections, and a rotor test tower aided research in many other areas. As usual, NACA researchers also pursued a multifaceted R&D program touching many other aspects of flight.

All of this postwar aeronautical activity received respectful and enthusiastic attention from press and public. Although the phenomenon of flight continued to enjoy extensive press coverage, events in the late 1950s suddenly caused aviation to share the limelight with spaceflight.

Enter Astronautics

World War II left many legacies, including an impressive array of new technologies spawned by massive military R&D programs—atomic energy, radar, antibiotics, radio telemetry, computers, big rockets like the V-2, jet engines, and a roster of additional gadgetry. Whether they were applied in conflict or in peace, these technological bequests inevitably shaped the world's destiny during the remainder of the twentieth century and into the twenty-first. There were geopolitical legacies as well. Much of Europe and Asia lay in ruins. The war had shattered old empires, and many postwar economies tottered on the brink of collapse. On opposite sides of the world stood the United States and the Soviet Union. The newly emergent superpowers forged contesting alliance systems that enveloped the globe and led to the protracted international confrontation called the cold war. In this environment both nations quickly exploited the new postwar technologies.

The atomic bomb was the most obvious and most menacing legacy of World War II. Both superpowers sought the best strategic systems that could deliver the bomb across the intercontinental distances that separated them. Both nations built propeller-driven intercontinental bombers and invested heavily in sprawling airbases to support them. The intercontinental rocket held great theoretical promise but seemed much further down the technological road. Atomic bombs were bulky and heavy; a rocket to lift such a payload would be enormous in size and expense. The Soviet Union doggedly went ahead with attempts to build such rockets. The American military relied on a new generation of jet bombers but also pushed the development of smaller research and battlefield rockets. The army imported Wernher von Braun and the German engineers who had created the wartime V-2 rockets and set them to overseeing the refurbishing and launching of captured V-2s at White Sands, New Mexico. The von Braun team later transferred to Redstone Arsenal, Huntsville, Alabama, where it formed the core of the Army Ballistic Missile Agency (ABMA). With its contractor the Jet Propulsion Laboratory (JPL), the army developed a series of battlefield missiles known as Corporal, Sergeant, and Redstone. The navy designed and built the Viking research rockets. The freshly independent air force, created in 1947, started a family of cruise missiles, intending to deploy them as intercontinental weapons.

By 1951 the progress that had been made on the development of a thermonuclear bomb of smaller dimensions revived interest in the long-range ballistic missile. Two months before President Truman announced that the United States would develop the thermonuclear bomb, the air force contracted with Consolidated Vultee Aircraft Corporation (later Convair) to resume study and then to develop the Atlas intercontinental ballistic missile (ICBM), a project that had been dormant for four years. During the next four years, a trio of intermediate range weapons (the army's Jupiter, the navy's Polaris, and the air force's Thor) and a second-generation ICBM, the air force's Titan, had been added to the list of American missile projects. Against the backdrop of cold war tensions, all received top national priority.

By the mid-1950s NACA had modern research facilities that had cost a total of $300 million and a staff totaling seventy-two hundred. Given the atmosphere of rivalry inherent in the cold war between the United States and the USSR and the national priority given to military rocketry, NACA's sophisticated facilities inevitably became involved. With each passing year, NACA scientists and engineers had been expanding the agency's missile research in proportion to the old mission of aerodynamic research. Major NACA contributions to the military missile programs came in 1955–57. Materials research led by Robert R. Gilruth at Langley confirmed that ablation could control the intense heat generated by warheads and other bodies reentering Earth's atmosphere; H. Julian

Allen at Ames demonstrated the blunt-body shape as the most effective design for reentering bodies; and Alfred J. Eggers at Ames did significant work on the mechanics of ballistic reentry.

During the mid-1950s, America's young space program brimmed with promise and projects. As part of its participation in the forthcoming International Geophysical Year (IGY), scheduled for 1957–58, the United States confidently proposed to launch a small satellite into orbit around the Earth. After a spirited design competition between the National Academy of Sciences–navy proposal (Vanguard) and the ABMA-JPL candidate (Explorer), the navy design emerged as the winner in September 1955. The navy's proposal succeeded largely because it did not seem to interfere with the high-priority military missile programs, since it would use a new booster based on the Viking research rocket. Also, it possessed a better tracking system and seemed to promise more scientific growth potential. By 1957 Vanguard was readying its first test vehicles for firing. The USSR had also announced that it would have an IGY satellite; the significance of this facet of the space race clearly extended beyond boosters and payloads to issues of global prestige.

On the military front, space activity gained increased momentum. Powerful military missiles now moved toward the critical flight test phase. Satellite ideas proliferated, though mostly on a sub rosa planning basis. Payload size and weight remained constant problems in all these concepts, given the limited thrust of the early rocket engines. Here the rapid advances in solid state electronics came to the rescue by reducing volume and weight; with new techniques such as printed circuitry and transistors, the design engineers could achieve new levels of miniaturization of equipment. Even so, heavier payloads inevitably loomed ahead; more powerful engines had to be developed. So design work began for several larger engines, topped by the monster F-1 engine, intended to produce eight times the power of the engines that lifted the Atlas, Thor, and Jupiter missiles.

All this activity still remained on the drawing board, work bench, or test stand on 4 October 1957, when the "beep, beep" signal from *Sputnik 1* echoed in electronic receivers around the world. The Soviet Union—not the United States—had orbited the world's first manmade satellite.

The American public's response seemed to be compounded of equal parts of alarm and chagrin. Americans had always believed that they led the world in modern technology. *Sputnik 1* rudely shattered that assumption. Not only had the Russians been first, but its satellite weighed an impressive 183 pounds, compared to Vanguard's intended 3 pounds, working up to 22 pounds in later

Test of Vanguard launch vehicle for U.S. International Geophysical Year (IGY) program to place satellites in Earth orbit to determine atmospheric density and conduct geodetic measurements. Malfunction in the first stage caused the vehicle to lose thrust after two seconds, and the vehicle was destroyed.

The three men responsible for the success of America's first Earth satellite, which was launched 31 January 1958. At the *left* is Dr. William H. Pickering, former director of JPL, which built and operated the satellite. Dr. James A. Van Allen, *center*, of the University of Iowa, designed and built the instrument on Explorer that discovered the radiation belts that circle Earth. At the *right* is Dr. Wernher von Braun, leader of the army's Redstone Arsenal team, which built the first-stage Redstone rocket that launched *Explorer 1*.

satellites. In the cold war environment, the contrast carried undefined but ominous military implications in terms of range and military payloads.

Fuel for such apprehensions accumulated rapidly. Less than a month after *Sputnik 1,* the Russians launched *Sputnik 2,* weighing a hefty 1,100 pounds and carrying a dog as passenger. President Eisenhower, trying to dampen the growing concern, assured the public of as yet undemonstrated American progress and denied that there was any military threat in the Soviet space achievements. As a counter, the White House announced the impending launch in December of the first Vanguard test vehicle capable of orbit. The government also belatedly authorized von Braun's army research team in Huntsville to plan a trial launch of their Explorer-Jupiter combination vehicle.

The media ballyhooed the carefully qualified announcement on Vanguard into great expectations of America's vindication. On 25 November Lyndon B. Johnson, Senate majority leader, chaired the first meeting of the Preparedness Investigation Subcommittee of the Senate Armed Services Committee, charged to review the whole spectrum of American defense and space programs.

Still the toboggan careened downhill. On 6 December 1957, the much touted Vanguard test vehicle rose about three feet from the launch platform, shuddered, and collapsed in flames. Its tiny three-pound payload broke away and lay at the edge of the inferno, beeping impotently.

Clouds of gloom deepened as the New Year began. Then, finally, a small rift appeared. On 31 January 1958, an American satellite at last went into orbit. Not Vanguard but the ABMA-JPL Explorer had redeemed American honor. True, the payload weighed only 2 pounds against the 1,100 of *Sputnik 2.* But the American launch achieved a scientific first: an experiment aboard the satellite reported the mysterious saturation of its radiation counters at 594 miles altitude. Professor James A. van Allen, the scientist who had built the experiment, thought this suggested the existence of a dense belt of radiation around Earth at that altitude. American confidence perked up again on 17 March when *Vanguard 1* joined *Explorer 1* in orbit.

Meanwhile, in these same tense months, both consensus and competition had been forming on the political front. An emerging consensus supported an augmented national space program as essential for the national interest. But competition over who would run such a program soon escalated. The Department of Defense (DoD) and the Atomic Energy Commission, dealing with rockets and nuclear systems, had congressional support. In the background, NACA began to build its own case as the manager for a bold American thrust into space.

NACA had devoted more and more of its facilities, budget, and expertise to missile research in the mid- and late 1950s. Under the skillful leadership of James H. Doolittle, chairman, and Hugh L. Dryden, director, the strong NACA research team had come up with a solid long-term, scientifically based proposal for a blend of aeronautic and space research. Also NACA offered reassuring experience of long, close working relationships with the military services in solving their research problems, while at the same time translating the research into civil applications. But NACA's greatest political asset lay in its peaceful, research-oriented image. President Eisenhower, Senator Johnson, and others in Congress all agreed that they wanted above all to avoid projecting cold war tensions into the new arena of outer space. Nevertheless, the shadow of the cold war and a sense of international competition hung in the background.

By March 1958 the consensus in Washington had jelled. The administration position (largely credited to James R. Killian, in the new post of president's special assistant for science and technology), the findings of Johnson's Senate subcommittee, and the NACA proposal converged. America needed a national space program. The military component would of course be under DoD. But a civil component, lodged in a new agency, technologically and scientifically based, would pick up certain of the existing space projects and forge an expanded program of space exploration in close concert with the military. After the requisite congressional hearings and debate, on 29 July 1958 President Eisenhower signed into law P.L. 85-568, the National Aeronautics and Space Act of 1958.

The act established a broad charter for civilian aeronautical and space research with unique requirements for dissemination of information, absorbed the existing NACA into the new organization as its nucleus, and empowered broad transfers from other government programs. The National Aeronautics and Space Administration came into being on 1 October 1958.

All of this made for a very busy spring and summer for the people in the small NACA Headquarters in Washington. Once the general outlines of the new organization were clear, both a space program and a new organization had to be charted. In August President Eisenhower nominated T. Keith Glennan, president of Case Institute of Technology and former commissioner of the Atomic Energy Commission, to be the first administrator of the new organization, NASA, and Dryden to be deputy administrator. Quickly confirmed by the Senate, they took the oath of office on 19 August. Talks with the Advanced Research Projects Agency identified the military space programs with characteristics that made them obvious candidates for transfer to the new agency.

Planners also began formulating the requirements for building a new center for space science research, satellite development, flight operations, and tracking. They eventually chose a five-hundred-acre site carved out of the Department of Agriculture's research center in Beltsville, Maryland. Construction began immediately, spurred by the growing sense of competition with the Soviet Union. In about two and one-half years, dignitaries arrived to dedicate the Robert H. Goddard Spaceflight Center (named for America's rocket pioneer) in March 1961. Along with this impressive new facility, committed solely to the goal of spaceflight, NASA acquired additional projects and physical assets in the name of space exploration. The agency's leaders now had to accomplish the effective integration of this new empire, as well as continue the research in aeronautics that had given such luster to NACA's—now NASA's—image.

Chapter 4

On the Fringes of Space, 1958–1964

On 1 October 1958, the 170 people in Headquarters gathered in the courtyard of their building, the Dolly Madison House, to hear Glennan proclaim the end of the forty-three-year-old NACA and the beginning of NASA. The new organization embraced 8,000 people, three laboratories (now renamed research centers), and two stations, with a total facilities value of $300 million and an annual budget of $100 million. On the same day, by executive order the president transferred additional assets to NASA: Project Vanguard and its 150-person staff and remaining budget from the Naval Research Laboratory; lunar probes from the army; lunar probes and rocket engine programs, including the F-1, from the air force; and a total of over $100 million of unexpended funds. NASA immediately delegated operational control of these projects back to the DoD agencies while it put its own house in order.

NASA did not live by space alone. In spite of the considerable attention given the nation's fledgling space program, the legacy of NACA's role in cutting-edge flight research continued to represent a lively field of endeavor. Subsequent years witnessed striking advances in hypersonic flight, epitomized by the remarkable series of X-15 aircraft, which generated equally dramatic strides in aeronautical technology. In addition, a roster of projects involving both military planes and civilian aviation yielded valuable results.

Toward Hypersonic Flight

When NACA set up the Muroc Flight Test Unit in 1948, Walter C. Williams began a decade of administration that saw many dramatic changes in the shapes and speeds of aircraft. The Muroc site won independence from Langley when it became the High-Speed Flight Research Station in 1954. Williams always argued for even more independence in the form of laboratory status, which would not only boost morale but also give the station greater prestige and autonomy. When NASA was created and the existing NACA labs were renamed as centers, old Muroc hands witnessed another change in names, becoming the NASA Flight Research Center (FRC) in 1959. Williams had to savor the change in names from a distance, since he already had been posted back to Langley as operations director for Project Mercury. But he could take pleasure in FRC's rapid growth and fame during the early 1960s, which was due largely to the test program for the X-15, a remarkably productive aircraft. After winning major headlines at the start of its flight tests, the X-15's success became eclipsed by NASA's space program. This was ironic, since the X-15 contributed heavily to research in spaceflight as well as to high-speed aircraft research.

The X-15 series were thoroughbreds, capable of speeds up to Mach 6.72 (4,534 MPH) at altitudes up to 354,200 feet (67 miles). There was a familiar European thread in the design's genesis. In the late 1930s and during World War II, German scientists Eugen Sanger and Irene Bredt developed studies for a rocket plane that could be boosted to an Earth orbit and then glide back to land. The idea reshaped American thinking about hypersonic vehicles. "Professor Sanger's pioneering studies of long-range rocket-propelled aircraft had a strong influence on the thinking which led to initiation of the X-15 program," NACA researcher John Becker wrote. "Until the Sanger and Bredt paper became available to us after the war we had thought of hypersonic flight only as a domain for missiles." A series of subsequent studies in America "provided the background from which the X-15 proposal emerged."

Momentum for such a plane gathered in 1951, when Robert Woods, the X-1 veteran from Bell Aircraft, proposed a Mach 5 research plane. By 1954 NACA accepted the hypersonic aircraft proposal as a major commitment. In the autumn of that year, NACA realized that it lacked funds to support the idea and joined forces with the air force and the navy; a memorandum of understanding gave NACA technical control of the effort, including flight testing and test reports. There was an undertone of military necessity in the memorandum, which declared that "accomplishment of this project is a matter of national urgency." The specifications and configurations circulated among potential bidders followed a pattern originally developed by a Langley team led by John Becker. "The proposals that we got back looked pretty much like the one we had put in," he recalled. NACA had certainly come a long way from testing aircraft designed and built by others. The earlier X-1 was something of a transition, involving Bell and NACA engineers. Although NACA in essence bootstrapped air force and navy funds for the X-15, it was very much a NACA idea and design from start to finish. In many ways, the X-15 program signaled a shift to the research, development, and management functions that characterized the NASA organization soon to come.

In the autumn of 1955, North American emerged as the winning contractor. Aside from building the plane, NACA and the armed services soon realized that they also had to develop other elements of a new system to support flight tests of the exotic X-15. The program called for fabrication of three research planes and a powerful new rocket engine to power them. The engine, a Thiokol XLR-99, had to be "man-rated" for repeated flights in the piloted rocket plane. For pilot training and familiarization, it was necessary to design and build a motion simulator and associated analog computer equipment. Before undertaking a 10- to 12-minute mission in the X-15, pilots spent a total of 8 to 10 hours practicing each moment of the test flight. Because of the extreme altitudes planned for X-15 missions, technicians needed to develop a unique, full-pressure flight suit. The planners had to lay out a special aerodynamic test range to monitor the X-15 as the plane streaked back to Edwards Air Force Base for its landing. The test range, officially labeled the High Altitude Continuous Tracking Radar Range, became known as the High Range. Following the X-15 program, the High Range continued to be an asset to flight testing of succeeding generations of aircraft.

The first X-15 arrived in the autumn of 1958, although powered flight tests did not start until September of 1959. In contrast to the secrecy surrounding the P-59 and the X-1, the X-15 program became a highly visible media event. In the wake of Sputnik, anything that seemed to redeem America's tarnished prestige in the "space race" automatically occupied center stage. Journalists flocked to Edwards for photos and interviews; Hollywood cranked out a hackneyed film about terse, steely-eyed test pilots and the rocket powered ships they flew. When the Mercury, Gemini, and Apollo programs began, the journalists migrated to hotter headlines in Florida. The X-15, meanwhile, moved into the most productive phase of its program, contributing to astronautics as well as aeronautics.

Between 1959 and 1968, the trio of X-15 aircraft completed 199 test flights. The fallout was far-reaching in numerous crucial areas, such as in hypersonic aerodynamics and in structures. During a test series to investigate high-temperature phenomena in hypersonic flight, temperatures on the surface soared to 1,300°F, so that large sections of the aircraft glowed a cherry-red color. The X-15's survival encouraged extensive use of comparatively exotic alloys, such as titanium and Inconel-X, which led to machining and production techniques that became standard in the aerospace industry. Although the cockpit was pressurized, the chance of accidental loss of pressurization in the near-space environment where the X-15 flew prompted development of the first practical full-pressure suit for pilot protection in space. The X-15 was the first to use reaction controls for attitude control in space; reentry techniques and related technology also contributed to the space program, and even earth science experiments were carried out by the X-15 in some of its flights.

The high-speed, high-altitude X-15, like the X-1, might be remembered as the epitome of an era, although NACA/NASA research activities, as usual, continued along many paths. For example, in the course of studies for supersonic cruise aircraft, two different trends of study began to emerge: a multimission combat plane operating at both high and low speeds and configurations for a supersonic transport.

The multimission plane idea took shape as a combat aircraft capable of sustained high speeds at high altitudes, as well as high speeds "down on the deck." This meant swept wings, which also decreased controllability and combat load at takeoff—unless the wings could be pivoted forward during takeoff and landing and swept back during flight. Test articles from wartime German experiments again pointed the way, and the Bell X-5 provided additional data during the early 1950s. The British also had a variable-sweep concept plane called the Swallow, which underwent extensive testing at Langley. The NASA contribution in this development included variable in-flight sweeping of the wings and the decision to locate the pivot points outboard on the wings rather than to pivot the wings on the centerline, solving a serious instability problem. All of this eventually led to the TFX program, which became the F-111. It became a long and controversial program, but the success of the variable-geometry wing on the F-111 and the navy's Grumman F-14 Tomcat owed much to NASA experimental work. The process of refining Mach 2 aircraft like these also led to profitable studies involving air inlets, exhaust nozzles, and overall drag reduction—factors that the aerospace industry applied to the new stable of Mach 2 combat planes in the following decades.

The work in high-speed combat planes paralleled growing interest in a supersonic transport. In 1959 a delegation from Langley briefed E. R. Quesada, head of the Federal Aviation Administration (FAA), on the technical feasibil-

The X-15 streaks across the western desert on a test run. Capable of flying at 6.7 times the speed of sound at altitudes exceeding 350,000 feet, the X-15 helped advance many aeronautical and spaceflight systems.

ity of a supersonic transport (SST). The NASA group advocated a variable-geometry wing and an advanced, fan-jet propulsion system. The briefing, later published as NASA Technical Note D-423, "The Supersonic Transport—A Technical Summary," analyzed structures, noise, runways and braking, traffic control, and other issues related to SST operations on a regular basis. An SST, the report concluded, was entirely feasible. The FAA concurred, and within a year a joint program with NASA had allocated contracts for engineering component development. Eventually, the availability of advanced air force aircraft provided the opportunity to conduct flight experiments as well. The idea of commercial airliners flashing around the globe at supersonic speeds received press attention, but the biggest headlines went to even more sensational developments in space, where human beings were preparing for inaugural voyages.

Getting into Space

Charged with a new space program, NASA scrambled to get this demanding venture up and running. During the first two years (1958–60) the agency pressed ahead to formulate a plan for its future in space, integrate its legacies, and fashion a new administrative structure. Only one week after NASA received its charter, Glennan gave the go-ahead to Project Mercury, America's first manned spaceflight program. The Space Task Group, headed by Robert R. Gilruth, established itself at Langley and immediately went to work. Excited by the new vistas of astronautics, many space-minded specialists signed on with NASA, where the need to meld diverse space projects into the prior nucleus of NACA procedures necessitated long hours of committee meetings and countless position papers. Given the urgent need for long-range planning, NASA's first ten-year plan received high priority, and copies arrived on Capitol Hill for congressional approval in February 1960. NASA's proposal called for an expanding program on a broad front: manned flight (first orbital, then circumlunar); scientific satellites to measure radiation and other features of the near-space environment; lunar probes to measure the lunar space environment and to photograph the Moon; planetary probes to measure and photograph Mars and Venus; weather satellites to improve our knowledge of Earth's broad weather patterns; and the development of larger, brawny vehicles for lifting heavier payloads. All of this implied intimidating levels of financial conundrums for an agency accustomed to annual spending levels of less than $100 million. Budgetary gurus estimated the costs of the space program at $1 billion to $1.5 billion per year over the decade ahead.

To conduct its space program, NASA obviously needed capabilities that it did not have. For one thing, it lacked booster vehicles to carry its proposed payloads into space. To that end Glennan sought to acquire the successful army team that had launched America's first satellite, the ABMA at Huntsville, Alabama, and its contractor, JPL, in Pasadena, California. Following some bureaucratic skirmishes, the army gave up JPL, and it was officially transferred 31 December 1958. Eventually the army also yielded ABMA; on 15 March 1960 ABMA's four-thousand-person Development Operations Division, headed by Wernher von Braun, went to NASA along with the big Saturn booster project. NASA later officially named this new organization the George C. Marshall Spaceflight Center (MSFC), after the revered World War II general, postwar secretary of state, and key figure in the Marshall Plan to revitalize postwar Europe.

As NASA's ten-year space exploration plan took shape and the capability grew, many other gaps needed attention. NASA's operational style differed markedly from NACA's in two important ways. First, NASA was going to be operational as well as do research. It would not only design and build launch vehicles and satellites, but it would launch them, operate them, track them, acquire data from them, and interpret the data. Second, it would do the greater part of its work by contract rather than in-house as NACA had done. The first of these required tracking sites in many countries around the world, as well as construction of facilities: antennae, telemetry equipment, computers, radio and landline communications networks, and so on. The second required the development of a larger and more sophisticated contracting operation than NACA had needed. In the first years, NASA leaned heavily on the DoD procurement system.

The problem of launch vehicles occupied much attention in these first years. A family of existing and future launch vehicles had to be structured for the kinds of missions and spacecraft enumerated in the plan. As much as NASA required larger boosters, an even more immediate problem involved the need to improve the reliability of existing hardware. By December 1959 the United States had attempted thirty-seven satellite launches; less than one-third had attained orbit. Electrical components, valves, turbopumps, welds, materials, structures—virtually everything that went into the intricate mechanism called a booster—had to be redesigned or strengthened or improved to withstand the stresses of launch. A new order of perfection in manufacturing and assembly had to be instilled in workers and managers. Rigorous, repeated testing had to verify each component, then subassembly, then the total vehicle. That bugaboo of the engineering profession, constant fiddling and changing in search of perfection, had to be constrained in the interest of reliability. Documentation seemed to mushroom. And since the existing vehicles originated from DoD products, NASA had to persuade DoD to enforce these rigorous standards on its contractors.

This was only one of the areas in which close coordination between NASA and DoD was essential and effective. In manned spaceflight, for example, several cooperative studies remained under consideration. Two depended on research aircraft like the X-15 as well as ballistic missile boosters like the Mercury Project. A third approach, the air force's Dyna-Soar, involved designs for a winged capsule to glide in stages down through Earth's atmosphere after a rocket boost into a shallow orbit. A fourth concept relied on a bathtub shape with stubby wings for a more direct reentry and landing.

In other areas, both NASA and DoD engaged in similar space scenarios. In the case of communications satellites, the military services needed "secure" units in orbit. NASA also planned orbiting units such as passive communications satellites, which "bounced" radar signals along as a limited communications relay system. Eventually, NASA received presidential approval to provide reimbursable launches to industry and to undertake its own research into active communications satellites. During the early 1960s, NASA agreed to provide launch services for American Telephone and Telegraph satellites and became increasingly active on its own. *Tiros 1,* launched 1 April 1960, inaugurated weather satellite information; *Echo 1,* a passive communications satellite, was

The Grumman F-14 Tomcat, with wings swept forward, was a legacy of variable geometry studies. This Navy F-14 underwent tests at Dryden Flight Research Center in 1986 and 1987 in a program to explore laminar flow on variable-sweep aircraft at high subsonic speeds.

launched 12 August 1960. As a large, balloonlike sphere inflated in orbit, it provided a passive target for bouncing long-range communications from one point on Earth to another. Perhaps as important, millions of people saw the moving pinpoint of light in the night sky—a visible and awesome harbinger of the dawning space age.

In the late 1950s, politics preoccupied the space program. Although not a direct campaign issue in the presidential campaign, the space program found little reassurance of its priority as an expensive new item in the federal budget. After John F. Kennedy's narrow election as president, the uncertainty deepened. Jerome B. Wiesner, the president-elect's science adviser, chaired a committee which produced a report both critical of the space program's progress to date and skeptical of its future.

Suddenly, another spectacular Soviet spaceflight claimed international headlines. On 12 April 1961, Soviet Cosmonaut Yuri Gagarin rode *Vostok 1* into a 187-by-108-mile orbit of Earth. After one orbit he reentered the atmosphere and landed safely—the first human to fly in space. Gagarin joined that elite pantheon of individuals who were the first to do the undoable—the Wright brothers, Lindbergh, now Gagarin. It seemed a faint consolation on 6 May 1961, when Project Mercury essayed its first manned spaceflight. Astronaut Alan B. Shepard Jr. rode a Redstone booster in his *Freedom 7* Mercury spacecraft for a fifteen-minute suborbital flight; parachutes floated his protective capsule down into the Atlantic; helicopters picked him out of the water some three hundred miles downrange.

The American record began to look somewhat better. But Gagarin had flown around the Earth, some 24,800 miles, compared with Shepard's 300. His Vostok weighed 10,428 pounds in orbit, contrasting with Mercury's 2,100 pounds in suborbit. Gagarin had experienced about 89 minutes in weightlessness, the mysterious zero-gravity condition that had supplanted the sound barrier as the great unknown. Shepard's mission involved only 5 minutes of weightlessness. By any unit of measure, the United States clearly struggled as a contender in spaceflight, especially in the indispensable prerequisite of rocket power. As the new president gloomily reflected: "We are behind . . . the news will be worse before it is better, and it will be some time before we catch up." The public reaction appeared less emphatic than after *Sputnik 1*, but congressional concern assumed a new intensity. Robert C. Seamans Jr., NASA's associate administrator and general manager, found himself hard put to restrain Congress from forcing more money on NASA than could be effectively used.

President Kennedy continued to be especially concerned. His inaugural address in January 1961 rang with an eloquent promise of bold new initiatives that would "get this country moving again." But the succeeding three months witnessed crushing setbacks—the Bay of Pigs invasion fiasco and the Gagarin flight. As one of several searches for new initiatives, the president asked his vice president, Lyndon B. Johnson, to head a study of what would be required in the space program to convincingly surpass the Soviets. Johnson, the only senior White House figure in the new administration with prior commitment to the space program, found strong support waiting in the wings. James E. Webb, the new administrator of NASA, had an established reputation as an aggressive manager of large enterprises, both in industry and during the Truman administration as director of the Bureau of the Budget and as undersecretary of state. Johnson's endorsement of space exploration came at a critical time. Backed by the seasoned technical judgment of Dryden, his deputy, and Seamans, his general manager, Webb moved vigorously to accelerate and expand the central elements of the NASA ten-year plan.

The largest single concept in that plan had been manned circumlunar flight. Now the question became, Could this country rally quickly enough to beat the

Soviets to that circumlunar goal? The considered technical estimate was, "Not for sure." But if we went one large step further and escalated the commitment to manned lunar landing and return, it became a new ball game. Both nations would have to design and construct a whole new family of boosters and spacecraft; this would be an equalizer in terms of challenge to both nations. The experts reporting to Webb seemed confident that the depth and competence of the American government-industry-university team would succeed.

Webb and his advisers were not content with a one-shot objective. The goal, they said, had to encompass a major space advance on a broad front—not just manned spaceflight but also a variety of boosters, communications satellites, meteorological satellites, and a commitment to planetary exploration.

This was the combined proposal presented to Johnson and approved and transmitted by him to Kennedy. It seemed the best new initiative the president had seen. So it was that on 25 May 1961 the president stood before a joint session of Congress and proposed a historic national goal:

> Now it is time to take longer strides—time for a great new American enterprise—time for this nation to take a clearly leading role in space achievement, which in many ways may hold the key to our future on earth. . . . I believe that this nation should commit itself to achieving the goal, before this decade is out, of landing a man on the moon and returning him safely to the earth. No single space project in this period will be more impressive to mankind, or more important for the long-range exploration of space; and none will be so difficult or expensive to accomplish.

The president correctly assessed the national mood. The challenge attracted widespread editorial support. Given the size of the commitment, congressional debate seemed almost perfunctory. Virtually without dissent, the decision to land an American on the Moon received official endorsement.

The Lunar Commitment

NASA felt exhilarated but awed. Dryden returned from a White House meeting to tell his staff that "this man" (Webb) had sold the president on landing a man on the Moon. Gilruth, immersed in what seemed to be big enough problems in the relatively modest Project Mercury, was temporarily aghast. But the die was cast for the nation's largest technological enterprise, dwarfing even the wartime Manhattan Project for developing the atomic bomb and the postwar crash development of strategic missiles.

NASA now had a blank check to go the Moon but remained unclear about

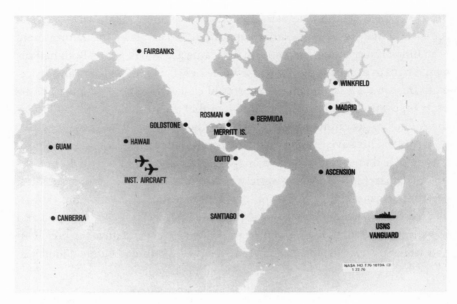

The worldwide satellite tracking network, 1976.

specifics. Dating from 1958, studies on circumlunar manned flight established a broad confidence that no major technological or scientific breakthroughs were needed to get a man to the Moon or even to land and return him. But a nagging number of precise operational unknowns persisted; the blank check caused them suddenly to loom larger.

Some points were clear. The massiveness of the effort would make this program different in kind from anything NASA had attempted. New organizational modes were essential; no one center could handle this program. A much stronger Headquarters team would be needed, coordinating the efforts of several centers and riding herd on an enormous mobilization of American industry and university effort.

Also, there were long-lead-time problems that demanded immediate attention, irrespective of later decisions. The solution to one of these problems was already three years under way—a big engine. Work on the 1.5-million-pound-thrust F-1 engine would be accelerated. Another difficulty had to do with the navigational challenge; accurate vectoring of a spacecraft from Earth to a precise point on the rapidly moving Moon 230,000 miles away presented a formidable problem in celestial mechanics. Therefore, the first large Apollo contract went to the Massachusetts Institute of Technology and its Instrumentation Laboratory, headed by C. Stark Draper, to begin study of this inscrutable problem and to develop the requisite navigational system.

The basic spacecraft could be delineated—one in which a crew would depart Earth, travel to the Moon, and return. It should have a baggage car, expendable oxygen, a jettisonable service module housing its propulsion, and other equipment. A whole new logistics system had to be fashioned; from factory to launch, everything threatened to outstrip normal sizes and normal transportation. There would have to be new factories, mammoth test stands, and huge launch complexes. Railroads and highways could not handle the larger components. Transportation by sea seemed to be the only answer, barging oversized stages down the Pacific coast, through the Panama Canal, and along the intracoastal canal around the Gulf of Mexico to Florida for launch. Other stages floated down the Tennessee and the Mississippi to the Gulf. A massive facility design and site location program had to begin even before the final configuration of the vehicle was decided. Limited in the facilities and construction area, NASA decided to call on the tested resource of the Army Corps of Engineers. It proved to be one of the wiser decisions in this hectic period.

As planning went forward in 1961 and 1962, order gradually emerged. A new concept for how to get to the Moon painfully surfaced: lunar-orbit rendezvous. A small group at Langley, headed by John C. Houbolt, had studied the trade-offs of direct ascent, Earth-orbit rendezvous, and other possibilities. They had been increasingly struck with the vehicle and fuel economics of this mission profile: after stabilizing in Earth orbit, a set of spacecraft would orbit around the Moon and, leaving the mother spacecraft in lunar orbit, dispatch a smaller craft to land on the lunar surface, reconnoiter, and rejoin the mother craft in lunar orbit for the return to Earth. Over a period of two years they refined their complex mathematics and argued their case. As time became critical for definition of the launch vehicle, they spoke before one NASA audience after another. Finally Houbolt, in a bold move, went outside of "channels" and got the personal attention of Seamans. This was an issue of such importance to the total program that an imposed decision was not enough; the major elements of NASA had to be won over and concur in the final technical judgment. Dismissed at first as risky and very literally "far out," lunar orbit rendezvous gradually won adherents. In July 1962 D. Brainerd Holmes, NASA director of manned spaceflight, briefed the House space committee on lunar orbit rendezvous, the chosen method of going to the Moon.

Once made, this decision permitted rapid definition of the Apollo spacecraft combination. The configuration of the launch vehicle had been determined

The original Mercury Seven astronauts with a U.S. Air Force F-106B jet aircraft. *Left to right:* M. Scott Carpenter, Leroy Gordon Cooper, John H. Glenn Jr., Virgil I. "Gus" Grissom Jr., Walter M. "Wally" Schirra Jr., Alan B. Shepard Jr., and Donald K. "Deke" Slayton.

seven months earlier. The objective envisioned a booster combination to carry a payload of nearly 300,000 pounds in Earth orbit and place 100,000 pounds in orbit around the Moon. To do this required a three-stage vehicle, the first stage employing the F-1 engine in a cluster of five, to provide 7.5 million pounds of thrust at launch. The second stage would cluster five new 225,000-pound-thrust liquid hydrogen and liquid oxygen engines (the J-2). The third stage, powered by a single J-2 engine, would boost the Apollo three-man spacecraft out of Earth orbit and into the lunar gravitational field. At that point the residual three-spacecraft combination would take over: a command module housing the astronauts, a service module providing propulsion for maneuvers, and a two-man lunar module (LM) for landing on the Moon. The engine on the service module would ignite to slow the spacecraft enough to be captured into lunar orbit; the fragile LM would leave the mother craft and descend to land its two passengers on the Moon. After lunar reconnaissance, the astronauts would blast off in the top half of the LM to rejoin the mother craft in lunar orbit, and the service module would fire up for return to Earth.

NASA also decided to construct a smaller launch vehicle, later dubbed the Saturn IB, for preliminary tests of the Apollo spacecraft in Earth orbit. Even this partial fulfillment of the Apollo mission required a first stage with 1.5 million pounds of thrust and a high-energy liquid oxygen–liquid hydrogen second stage.

But in the articulation of this grand design, vast gaps in experience and technology became evident. At three critical points, the master plan depended on successful rendezvous and docking of spacecraft. Although theoretically feasible, this had never been done and was not within the scope of Project Mercury. Short of an intricate, hideously expensive, and possibly disastrous series of tests with unproven Apollo hardware, how could NASA master the techniques of rendezvous and docking? Could astronauts function in space outside the spacecraft? A myriad of other questions emerged. Clearly something was needed between the first steps of Mercury and the grand design of Apollo. The gap appeared to be too great to jump when lives were at stake.

Even Mercury sometimes seemed impossible. But slowly, stubborn problems yielded. The second suborbital flight, *Liberty Bell 7*, lifted off on 21 July 1961; its sixteen-minute flight went well, though on landing the hatch blew off prematurely and the spacecraft sank, just after Astronaut Virgil I. Grissom was hoisted to safety in a rescue helicopter. In September the unmanned Mercury-Atlas combination completed a successful orbit and descended as planned, east of Bermuda. On 29 November the final test flight took chimpanzee Enos on a two-orbit ride and landed him in good health.

NASA officials cleared the system for manned orbital flight. And on 20 February 1962, Astronaut John H. Glenn Jr. became the first American to orbit Earth in space. *Friendship 7* circled Earth three times; Glenn flew parts of the last two orbits manually because of trouble with his autopilot.

The United States took its astronaut heroes to its heart with an enthusiasm that bewildered them and startled NASA. Their mail was enormous; hundreds of requests for personal appearances poured in. Glenn had a rainy parade in Washington and addressed a joint session of Congress. On 1 March 4 million people in New York showered confetti and ticker tape on him and fellow astronauts Shepard and Grissom. Nor was the event unnoticed by the competition. President Kennedy announced the day after the Glenn flight that Soviet Premier Nikita Khrushchev had congratulated the nation on its achievement and had suggested that the two nations "could work together in the exploration of space." The results of this exchange were a series of talks between Dryden of NASA and Anatoliy A. Blagonravov of the Soviet Academy of Sciences. By the end of the year, they had agreed to exchanges of meteorological and magnetic-field data and some communications experiments.

Additional feats made 1962 a big year for the young American space program. There were two more Mercury flights and successful flight tests for the big Saturn I booster. For AT&T, NASA launched the first active communications satellite, *Telstar I*. The Tiros meteorological satellites operated so well that the U.S. Weather Bureau regularly integrated Tiros data into its operational forecasting and planned its own weather satellite system. Booster reliability experienced major improvement.

Not all was sweetness and light, however. The Ranger, designed to photograph the moon as it dived into the lunar surface, experienced five straight failures, with a sixth in 1963. Congressional pressure forced a painful reorganization by NASA, JPL, and Hughes Aircraft, the contractor. The last three flights performed brilliantly, dispelling many of the scare theories about a dangerously soft lunar surface. More reassurance came with subsequent missions of Surveyor, the first lunar soft-lander, and Lunar Orbiter, a mapping satellite to orbit the Moon and snap pictures of potential landing sites for Apollo explorers.

Spin-off

As the dimensions of Apollo began to dawn on Congress and the scientific community, there were rumbles: Apollo would preempt too much of the scientific manpower of the nation; Apollo was an "otherworldly" stunt, directed at the Moon instead of at pressing problems on Earth. Administrator Webb met both of these caveats with positive programs.

In acknowledgment of the drain on scientific manpower, Webb won White House support for a broad program by NASA to augment the scientific manpower pool. Thousands of fellowships were offered for graduate study in space-related disciplines, intended to replace or at least supplement the kinds of talent engulfed by the space program. Complementing the fellowships was an even more innovative program, government-financed buildings and facilities on university campuses for the new kinds of interdisciplinary training that the space exploration effort required.

By the end of the program in 1970, NASA had paid for the graduate education of five thousand scientists and engineers; more than $250 million went to scholarships, laboratory construction, and research grants. The NASA university program proved very effective: on the political side, it reduced tensions between NASA and the scientific-engineering community; on the score of national technology capability, it enlarged and focused a large segment of the research capabilities of the universities.

To refute the other charge—that Apollo would serve only its own ends and

Launches of the Saturn I, shown here, and the similar Saturn IB increased NASA's confidence in engines, boosters, and spacecraft, preparing the way for eventual human missions to the Moon during Project Apollo.

Mercury, Gemini, and Apollo spacecraft all splashed down into the ocean and had to be retrieved by helicopter. The Sikorsky UH-34D lost its struggle with Gus Grissom's *Liberty Bell 7* capsule on 21 July 1961. Minutes after Grissom got out of the spacecraft, it sank.

not the broader needs of the nation's economy—Webb created the NASA Technology Utilization Program in 1962. Its basic purpose was to identify and hold up to the light the many items of space technology that could be or had been adapted for uses in the civilian economy. By 1973 some 30,000 such uses had been identified, and new ones were rolling in at the rate of 2,000 a year.

But the program went beyond that. A concerted effort was made in every NASA center not only to identify possible transfers of space technology but also to use NASA technical people and contractors to explore and even perform prototype research on promising applications. Every NASA center began to host an annual open house, where visitors and agency personnel looked for answers and traded ideas. New products ranged from quieter aircraft engines to microminiaturized and solid-state electronics that revolutionized TV sets, radios, and small electronic calculators. NASA's computer software programs enabled a wide range of manufacturers to test the life history of new systems; they could predict potential problems, how the systems would perform, how long they would last, and so on.

Many other facets of the space program assumed significance for the quality and sustenance of life for citizens of the planet. NASA's experience in microminiaturized electronics and in procedures to protect and monitor the health of astronauts during missions generated hundreds of medical gadgets and techniques that saved lives and enhanced health care. In the field of energy, NASA developed broad experience in devices to collect, store, transmit, and apply solar, nuclear, and chemical energy for production of mechanical and electrical power. One example was the growing use of solar cells; another involved compact, longer-life batteries.

The catalog of space benefits continued to grow. By the 1970s communications satellites made invaluable contributions to long-range communications. Real-time TV coverage brought coronations, Olympic Games, and global crises into the living rooms of families around the world. Satellite data transmissions became everyday functions for international businesses, banks, and airlines. Weather satellites became indispensable equipment in tracking hurricanes, providing advanced warning that protected property and saved lives. Satellites also became essential in long-range weather forecasting. In the United States and elsewhere around the globe, the space program affected the daily lives of millions, whether they were aware of it or not.

Gemini

Against the background of progress in aerospace R&D, several critical issues remained. For one thing, the vexing questions of rendezvous and extravehicular activity still had to be answered. So on 3 January 1962, NASA announced a new manned spaceflight project, Gemini. Using the basic configuration of the Mercury capsule, enlarged to hold a two-man crew, Gemini was to fit between Mercury and Apollo and provide early answers to assist the design work on Apollo. NASA arranged to use the powerful Titan II missile being developed by the air force as the launch vehicle. For a target vehicle with which Gemini could rendezvous, NASA chose the air force's Agena; launched by an Atlas, the second-stage Agena had a restartable engine that enabled it to have both passive and active roles. The same Space Task Group responsible for operating Mercury took on the additional assignment of managing the Gemini program.

Gemini began as a Mark II Mercury, a "quick and dirty" program. The only major engineering change aside from scale-up was to modularize the various electrical and control assemblies and place them outside the inner shell of the spacecraft to simplify maintenance. But perhaps no engineer alive could have left it at that. After all, Gemini was supposed to bridge to Apollo. Here was a chance to try out ideas. If they worked, they would be available for Apollo. Engineers designed the spacecraft to have more aerodynamic lift than Mercury, so the pilot could have more landing control; fuel cells (instead of batteries) with enough electric power to support longer-duration flights; and fighter plane–type ejection seats for crew abort, to supersede the launch escape rocket that perched on top of Mercury.

With the Mercury program and the spacecraft design role in Apollo, and now Gemini, it became clear that the Space Task Group needed a home of its own and some growing room. On 19 September 1961, Administrator Webb announced a new Manned Spacecraft Center, to be built on the outskirts of Houston. It would house the enlarged Space Task Group, now upgraded to a center, and would have operational control of all manned missions as well as being the developer of manned spacecraft. For transportation of oversized Apollo components, the location provided water access to the Gulf of Mexico through the ship channel to Galveston. Moreover, the Houston metropolitan area answered NASA requirements for skilled labor and technically oriented industry as a necessary infrastructure.

Water access played a role in all site selections for new Apollo facilities. The big Michoud Ordnance Plant outside New Orleans, where the ten-meter-diameter Saturn V first stage would be fabricated, resided on the Mississippi River; the Mississippi Test Facility, with its huge test stands for static firing tests of the booster stages, occupied a site along the Gulf of Mexico, in Pearl River County, Mississippi.

All this effort would come together at the launch site at Cape Canaveral, Florida, where NASA had a small Launch Operations Center, headed by Kurt H. Debus. NASA had been a tenant there, using air force launch facilities and tracking range. Now Apollo loomed. Apollo required physical facilities much too large to fit on the crowded Cape. For safety's sake there would have to be large buffer zones of land around the launch pads; if a catastrophic accident occurred, in which all stages of the huge launch vehicle exploded at once, the force of the detonation would approach that of a small atomic bomb. So NASA sought and received congressional approval to purchase more than 111,000 acres of Merritt Island, just northwest of the air force facilities. Lying between the Banana River and the Atlantic, populated mostly by orange growers and the ubiquitous alligators, Merritt Island had the requisite water access and safety factors.

Planners struggled through 1961 with a wide range of concepts and possibilities for the best launch system for Apollo, hampered by having only a gross

Kennedy Space Center as it appeared in the mid-1960s. The 350-foot-tall Saturn V launch vehicle emerged from the cavernous Vehicle Assembly Building aboard its crawler and began its stately processional to the launch complex three miles away.

knowledge of how the vehicle would be configured, what the missions would involve, and how frequent the launches would be. Finally on 21 July 1962, NASA announced its choice: the Advanced Saturn (later Saturn V) launch vehicle would be transported to the new Launch Operations Center on Merritt Island stage by stage; the stages would be erected and checked out in an enormous vehicle assembly building; the vehicle would be transported to one of the four launch pads several miles away by a huge tractor crawler. This system constituted a major departure from previous practice at the Cape; launch vehicles had usually been erected on the launch pad and checked out there. Under the new concept, the vehicle would be on the launch pad for a much shorter time, allowing for a higher launch rate and better protection against weather and salt spray. As with the other new Apollo facilities, NASA enlisted the Corps of Engineers to supervise the vast construction project.

The simultaneous building of facilities and hardware required a great deal of money and a great many skilled people. The NASA budget, $966.7 million in fiscal 1961, totaled $1.825 billion in 1962. It hit $3.674 billion the next year and by 1964 rose to $5.1 billion; it remained near that level for three more years. In personnel, NASA doubled in those same years from 17,471 to 35,860. Of course this looked like small potatoes compared to the mushrooming contractor and university force, where 90 percent of NASA's money was spent. When the Apollo production line peaked in 1967, more than 400,000 people worked on some aspect of Apollo.

As the large bills began to come in, some wincing occurred in the political system. President Kennedy wondered briefly if the goal was worth the cost; in 1963 Congress had its first real adversary debate on Apollo. Administrator Webb had to point out again and again that this was not a one-shot trip to the Moon but the building of a national space capability that would have many uses. He also needled congressmen with the fact that the Soviets were still ahead; in 1963 they were orbiting two-man spacecraft, flying a 129-mile-orbit tandem mission, and orbiting an unmanned prototype of a new spacecraft. Support rallied. The Senate rejected an amendment that would have cut the fiscal 1964 space budget by $500 million. The speech that President Kennedy planned to deliver in Dallas, Texas, on that fateful 22 November 1963 would have defended the expenditures of the space program. His successor, Lyndon Johnson, was committed to the goal of American leadership in space exploration.

As 1963 drew to a close, NASA could feel that it was on top of its job. Initiatives like the phenomenal X-15 plane kept it in the vanguard of high-speed research. The master plan for Apollo was drawn; the organization and the key people were in place.

Chapter 5

Dress Rehearsals, 1964–1969

The Gemini program, planned to launch two-man crews into Earth orbit for several days, played a significant role in verifying systems and hardware, acquiring operational experience, and streamlining management of the increasingly complex manned space venture. Although space science often took a back seat as the Apollo effort forged ahead, this field of research experienced solid gains. Satellite communications opened the door to a new era of transmitting news and information. Using wind tunnels and experimental aircraft—especially "lifting body" designs—NASA continued to probe various elements of aeronautics that not only supported subsequent aerospace initiatives but also accrued worthwhile data for later generations of aircraft.

Gemini's Legacy

Late in 1963 President Lyndon B. Johnson signed the paperwork that changed the name of the Launch Operations Center to the Kennedy Space Center. The untried Gemini capsule, which awaited launch early in 1964, became the first tenant under the new name. Anxious to make everything work perfectly on the first launch, the checkout crew's ministrations crawled through the winter into spring, when Gemini finally made an unmanned orbit that verified the vehicle-spacecraft combination of Titan II and the Gemini capsule. NASA wanted to fire off Gemini missions on a three-month cycle, but the qualification process of Gemini for manned operations seemed to be cursed. The second unmanned

Gemini capsule arrived late, got hit by lightning on the pad, experienced delays caused by hurricanes in the region, and went through an aborted launch when the Titan's engines ignited but then shut down. *Gemini 2* finally went into orbit in January 1965—hardly a three-month cycle—and confirmed all the guidelines necessary to make it man-rated.

In March the manned mission of *Gemini 3* got things back on schedule and turned in a confidence-building three orbits. Although the remainder of the Gemini program included some frustrating moments, they contributed significant operational experience for the Apollo effort that lay ahead. For one thing, Gemini pushed the orbital time in space from a few days to two weeks, giving NASA confidence that its crews and equipment could cope with the length of time needed to travel to the Moon, explore its surface, and return to Earth. Physiologically, astronauts on these extended missions experienced temporary effects such as slower heartbeats, some loss of calcium in their bones, and other manifestations of weightlessness, but they all returned to normal after a few days back on the ground.

Another crucial question for Apollo concerned the three rendezvous and docking maneuvers planned for every lunar flight. Nudging two spacecraft together while cruising along in space raised the anxiety level of all concerned. *Gemini 3* made the tentative beginning by testing the new thruster rockets with short-burst firings that changed the height and shape of orbit and by one maneuver that for the first time shifted the plane of the flight path of a spacecraft. *Gemini 4* tried to rejoin its discarded second-stage booster, but faulty techniques burned up too much maneuvering fuel, and the pursuit had to be abandoned. This sent engineers back to their computers to figure out some better ideas. *Gemini 5* tested out the techniques and verified the performance of the rendezvous radar and rendezvous display in the cockpit. The Gemini series also reminded NASA about the virtues of patience and improvisation.

During one subsequent mission, the loss of an Agena target stage to test rendezvous procedures sent mission planners scrambling for alternatives. Eventually, they substituted another pair of Gemini launch vehicles to carry out the rendezvous procedure. During the *Gemini 8* mission in March 1966, after docking with Agena, astronaut Neil Armstrong had to fire his reentry rockets to bring a tumbling Gemini spacecraft under control. Later missions encountered additional problems, until the flights of *Gemini 10* and *Gemini 11* confirmed rendezvous and docking techniques during Gemini operations later in the year. Another set of Gemini tasks included extravehicular activity (EVA) to determine if astronauts could perform useful work outside an orbiting spacecraft as well as when humans arrived on the Moon.

The view from *Gemini XI*'s window of the Agena rocket, with which the Gemini crew is practicing rendezvous and tethered station keeping.

In preparation for the strange floating sensation they expected to encounter in space, astronauts trained for the phenomenon by doing the same assignments in water tanks on Earth. The real thing turned out to be more problematic than NASA anticipated. *Gemini 4* began EVA when Edward H. White II floated outside his spacecraft for twenty-three minutes. Protected by his spacesuit and attached to Gemini by a twenty-five-foot umbilical cord, White used a hand-held maneuvering unit to move about, took photographs, and in general had such an exhilarating experience that he had to be ordered back into the spacecraft. Because he had no specific work tasks to perform, his EVA seemed deceptively easy. On subsequent missions, when work tasks proved difficult and EVA required extended time, astronauts overtaxed the spacesuit system or became exhausted. NASA learned its lesson. When *Gemini 12* went up, many additional body restraints, handholds, and footholds embellished the crew module.

Primarily, Gemini represented a technological learning experience. Consequently, of the 52 experiments in the program, more than half (27) were technological, exploring the limits of the equipment. But Gemini crews also carried out 17 scientific activities and 8 medical experiments. One important task accumulated fourteen hundred color photographs taken of Earth from various altitudes. This provided investigators the first large corpus of color photographs from which to learn more about the planet on which we live. Probably the most valuable management payoff from Gemini was the operational one: how to live and maneuver in space; next was how to handle a variety of situations in space by exploiting the versatility and depth of the vast NASA-contractor team that stood by during flights. Finally, there were valuable fiscal lessons: an advanced technology program had a "best path" between too slow and too fast. Deviation on either side, as in the early days of Gemini, could cost appalling amounts of money. But once on track, even economies were possible.

Gemini became a versatile, flexible spacecraft system that wound up exploring many more nooks and crannies of spaceflight than its originators ever foresaw. Major lessons were transmitted to Apollo: rendezvous, docking, EVA, manned flights up to two weeks in duration. Gemini advanced the state of the art in a multitude of hardware areas: thrusters, fuel cells, environmental control systems, space navigation, space suits, and ancillary equipment. Equally important, there now existed a big experience factor for the astronauts and for the people on the ground, in the control room, around the tracking network, and in industry. Gemini seasoned a total team that solved real-time problems in space with men's lives at stake. In the development stage of Apollo, the bank of knowledge from Gemini paid off in hundreds of subtle ways.

Boosters and Spacecraft for Apollo

Throughout Gemini's operational period, Apollo slogged along toward completed stages and completed spacecraft. As precursors to the towering Saturn V rocket envisioned for the lunar expeditions, NASA authorized two interim versions to verify the concept of clustered engines to lift the daunting payloads required for manned missions to the distant Moon. The Saturn I and its larger sibling, Saturn IB, both featured high-energy liquid hydrogen upper stages. At Marshall Spaceflight Center, a sprawling area of huge test stands and laboratories operated day and night to develop the powerful Saturn stages. Saturn I, the booster almost overtaken by events, finished its ten-flight program in 1964 and 1965 with six launches featuring a liquid-hydrogen second stage. Understanding how to handle such volatile propellants and manage their design and engineering for use with Apollo launch components and acquiring the confidence to rely on them in manned missions were significant steps forward in the Apollo program. The Saturn I series not only proved out the clustered-engine concept but also tested early models of the Apollo guidance systems.

On 3 June 1965 Edward H. White II became the first American to step outside his spacecraft and let go, effectively setting himself adrift in the zero gravity of space. For twenty-three minutes White floated and maneuvered himself around the Gemini spacecraft while logging 6,500 miles during his orbital outing.

NASA assigned operational status to the last four launches. One hoisted a roughly built test model (called a "boilerplate") of an Apollo spacecraft into orbit, and the final three flights carried Pegasus meteoroid-detection satellites into orbit. Their statistics reassured mission planners that spacecraft and astronaut crews could rest easy about the threat of disastrous space debris slamming into them during a mission. Last but not least, the final pair of Saturn I boosters had been fabricated entirely by NASA's industrial contractors (Chrysler for the first stage, and McDonnell Douglas for the second),* mark-

*Douglas Aircraft merged with McDonnell Corporation in 1967.

ing a transition from the army-arsenal in-house concept that had previously characterized the Marshall Spaceflight Center.

Meanwhile, the larger brother, Saturn IB, began to take shape. Its first stage could generate 1.6 million pounds of thrust from eight of the Rocketdyne H-1 engines that had powered Atlas and Saturn I, now uprated to 200,000 pounds each. The second stage featured a single engine, the new Rocketdyne J-2 liquid hydrogen propulsion system, generating 200,000 pounds of thrust. It formed a crucial element of the forthcoming Saturn V vehicle, since a five-engine cluster would power the second stage and a single J-2 would power the third stage. In its application on the third stage, engineers also designed the J-2 with the critical capability to restart in orbit, propelling the Apollo spacecraft on a translunar trajectory toward the Moon.

Saturn IB became the first launch vehicle to be affected by a new concept called "all-up" testing. Associate Administrator George Mueller, pressed by budgetary constraints and relying on his industry experience in the air force's Minuteman ballistic missile program, pressed NASA to abandon its stage-by-stage testing. With intensive ground testing of components, he argued, NASA could reasonably test the entire stack of stages in flight from the beginning, at great savings to budget and schedule. Having built their splendid success record through conservative, step-by-step increments, Marshall engineers vigorously opposed the new concept. But eventually Mueller triumphed. On 26 February 1966 the complete Saturn IB flew with the Apollo command and service module in suborbital flight; the payload was recovered in good condition. Two subsequent flights confirmed operational expectations for the instrument unit, the service module, the command module, and the heat shield for reentry at 25,000 MPH.

The biggest booster of all, Saturn V, consumed considerable management attention in getting it pieced together. Developed by three different contractors (Boeing for the first stage, North American—later Rockwell International—for the second, and McDonnell Douglas for the third), the three stages of Saturn V possessed individual histories and problems. The first stage, although the largest, had a long lead time and remained on schedule despite headaches caused by its cluster of five huge F-1 engines. The third stage, though enlarged and more sophisticated in comparison with the version flown on Saturn IB, benefited from its prior flight experience. The second stage, with five J-2 engines consuming liquid oxygen, gave NASA and its contractor the biggest headaches. The S-II stage, as it was called, endured persistent reengineering to shave weight and improve performance, because the other two stages had already progressed so far in the fabrication process. The S-II became the pac-

As the space program grew more complex, logistics became a major challenge. Beginning in the 1960's, oversized cargo was carried by the Super Guppy, a dramatically modified Boeing Stratoliner. The Aero Spacelines B377SGT Super Guppy was at Dryden in May 1976 to ferry the X-24 and HL-10 lifting bodies from Dryden to the Air Force Museum at Wright-Patterson Air Force Base, Ohio. The Super Guppy and the Super Guppy Turbine remained in occasional use through the 1990s.

ing item of the Saturn V and remained in this tenuous category almost until the first launch.

Of the three spacecraft, the LM prevailed, early and late, as the problem child. For one thing, engineering for it commenced late—a whole year late. For another, it differed radically from previous space equipment. The LM consisted of two discrete spacecraft; one to descend to the lunar surface from lunar orbit; the other to separate from the descent stage, leap off the lunar surface into lunar orbit, and rendezvous with the command module. The engine for each stage absolutely had to work perfectly for the one time it fired. Both experienced teething troubles. The descent engine became particularly troublesome, to the point where NASA let a second contract for a backup engine of different design. Weight persisted as a never-ending problem with the LM. Each small change in a system, each substitution of one material for another, had to be considered as much in terms of pounds added or saved as in any gain in system efficiency. All of this kept the engineers at Marshall Spaceflight Center guessing as well, since they had to make sure that the Saturn V had enough power to lift everything into orbit. As von Braun wryly remarked, "every time we talked to the Houston people, the damn LM had gotten heavier again." Not until the end of 1966 did the Saturn 1B and the block 1 Apollo command and service module win NASA's imprimatur as man-rated.

On 27 January 1967, *AS-204*, to be the first Apollo manned spaceflight, sat on the launch pad at Cape Kennedy, moving through preflight tests. After suiting up, astronauts Virgil I. Grissom, Edward H. White II, and Roger B. Chaffee settled into the command module, secured the entry hatch, and began proceeding through the countdown toward a simulated launch. At T-minus-10 minutes tragedy struck without warning. As Major General Samuel C. Phillips, Apollo program director, described it to news reporters, ground crew technicians heard a radio call from inside the capsule that fire had broken out. Before they could react, the stunned ground crew saw flames break through the spacecraft shell and envelop the spacecraft in smoke, Phillips said. Rescue attempts failed. It required a torturous five minutes to get the hatch open from the outside. Long before that, the three astronauts died from asphyxiation. The tragedy of *AS-204* became the first fatal accident in the American spaceflight program.

Shock swept across the nation and the world. In the White House, President Johnson had just presided over the signing of an international space law treaty when Administrator Webb phoned with the crushing news. In a public comment the next day, Webb admitted that he had accepted the inevitability of a deadly incident sometime during the space program. But he had never expected it to happen on the ground.

The day following the fire, Deputy Administrator Seamans appointed an eight-member review board to investigate the accident. On 10 April Webb and other NASA executives briefed the House space committee on the findings: the fire had apparently been started by an electrical short circuit, which ignited the oxygen-rich atmosphere and fed on combustible materials in the spacecraft. But the basic spacecraft design remained sound. Meanwhile, other accident teams waded through minute reviews of spacecraft design, wiring, combustible materials, test procedures, and dozens more items. Eighteen months and $50 million later, a substantially redesigned and reengineered block II spacecraft received design certification for upcoming Apollo missions. Nonetheless, after hearings in both houses, congressional ire lingered, gradually eroding Webb's support on Capitol Hill.

None of this made it any easier for NASA to garner support for its ambitious long-range plans in the exploration of space. Well before men flew in Apollo spacecraft, the question had been raised as to what, if anything, NASA proposed to do with humans in space after Apollo was over. With the long lead times and heavy costs inherent in manned space programs, advance planning was essential. As early as 1964, President Johnson raised the issue with Webb. Subsequent studies yielded reports that included lunar surface exploration operating out of an unmanned Apollo LM landed on the Moon, and Earth-orbital operations leading to space stations.

Spacecraft for Space Science

Manned spaceflight, with its overwhelming priority, demonstrated both direct and indirect impact on the NASA space science program. From 1958 to 1963, scientific satellites recorded impressive discoveries: the van Allen radiation belts, Earth's magnetosphere, the existence of the solar wind. In the next four years, investigators directed much of the space science effort toward finding more detailed data on these extensive phenomena. The radiation belts were found to be indeed plural, with definite, if shifting, altitudes. The magnetosphere was found to have an elongated tail that reaches out beyond the Moon and through which the Moon periodically passes. The solar wind was shown to vary greatly in intensity with solar activity.

All of these stood out as momentous discoveries about our nearby space environment and hinted of tantalizing cosmic mysteries in deep space. So NASA helped plan and develop a second generation of spacecraft called the observatory class. Five to ten times as heavy as early satellites, they carried designations such as solar observatories, astronomical observatories, and geophysical observatories. As these complex spacecraft arrived in space during the mid-1960s, the first results seemed somewhat disappointing. Fleeting results confirmed their promise, but their very complexity inflicted them with short lifetimes and electrical failures. Their champions in the scientific community still held solid expectations that the assorted glitches could be worked out for subsequent launches. But by the late 1960s the impingement of manned spaceflight budgets on space science budgets reduced or eliminated many of these promising starts.

Because so many lunar missions featured elements applicable to the Apollo program, NASA managers allowed most to run their course. A series of Surveyor probes made lunar landings and confirmed that the Moon's surface, described by news releases as having the consistency of wet sand, would not crumble when a manned LM set down. The Lunar Orbiter project produced invaluable maps, including potential landing sites. Despite heavy cuts in planetary programs, the first two flights of the Mariner series provided exciting new glimpses into the history of the solar system. *Mariner 4* sailed past Mars on 14

The wingless lifting body aircraft are sitting on Rogers Dry Lake at what is now NASA's Dryden Flight Research Center, Edwards, California. *Left to right:* the X-24A, the M2-F3, and the HL-10. The lifting body aircraft were used to study the feasibility of maneuvering and landing an aerodynamic craft designed for reentry from space. These lifting bodies were air launched by a B-52 mother ship, then flew powered by their own rocket engines before making an unpowered approach and landing.

July 1965 and produced the first close-up view of Earth's fabled neighbor. But the immediate future of more sophisticated planetary exploration eventually faded into oblivion.

In a different context, the saga of applications satellites geared toward practical duties fared well. The applications satellites were a crowning achievement for NASA in the early 1960s. The NASA policy of bringing a satellite system along through the R&D stages to flight demonstration of the system and then turning it over to someone else to convert into an operational system received its acid test in 1962. With the demonstration of Syncom performance, the commercial potential of communications satellites became obvious and immediate. NASA's R&D role seemed over, but controversy ensued over the issue of transferring Syncom's commercial potential to private ownership while avoiding the pitfall of favoritism. Eventually, the Kennedy administration proposed to authorize the Communications Satellite Corporation, a unique government-industry-international combination. Overcoming Senate intransigence, including a filibuster, the administration proposal became law in 1962, and ComSat went into business.

On a reimbursable basis, NASA provided launch services, and three years later ComSat's first communication satellite, *Early Bird I*, went into orbit. The same year, Russia inaugurated a competitive system of Molniya satellites with partners from several Iron Curtain countries and from France. In the meantime, the United States and ComSat collaborated in the evolution of an international network, Intelsat, with five advanced communications satellites in synchronous orbits, some 20 of an expected 40 ground stations in operation, and forty-eight member nations participating. On the American side, the question of government-sponsored research on communications satellites did not go away after the creation of ComSat. Congress continued to worry over the thorny question of whether the government should carry on advanced research on communications satellites if the only result was a permanent, government-sponsored monopoly.

For NASA, weather satellites seemed simpler in the sense that the relationship involved only two U.S. government agencies. The Weather Bureau seized on the highly successful *Tiros* as the model for its operational satellite series. Navigational satellites, one of the bright, early possibilities of space, continued to be intractable. But a new entry appeared, the earth resources satellite. Impressed by the *Tiros* photographs and even more by the Gemini photographs, the Department of Interior suggested an earth resources satellite program in 1966. Early NASA investigation envisioned a small, low-altitude satellite in Sun-synchronous orbit. The earth resources initiatives marked a new approach for NASA, involving many more government agencies and many more political minefields than the comparatively uncluttered researches in space.

Aspects of Flight Research

In the world of aviation, the advanced research activities of NASA also became more subtle and difficult to track. An interlocking network of basic research, applied research, and theoretical research evolved to feed new ideas and options into the planning process. The most visible portion involved flight research. Like many other NASA operations, it sometimes wound up supporting work in the space program.

Although ballistic reentry from space had become familiar by the 1960s, one group of engineers doggedly argued in favor of "lifting" reentry, an idea that dated from NACA era. The challenge was to build a spacecraft with aerodynamic characteristics so that a crew could fly back through Earth's atmosphere and land at an airfield. The X-20A Dyna-Soar proposed by the air force was one such example.

But the Dyna-Soar, a victim of budget constraints and new technology of the 1950s, never flew. NACA became involved in a smaller series of lifting body aircraft that helped pave the way for the space shuttle design. At Ames a series of exploratory studies during the 1950s culminated in a design known as the M2, a modified half-cone (it was flat on the top) and a rounded nose to reduce heating. NASA engineers at Edwards kept up with many of the theoretical ideas percolating out of Ames, and Robert Reed, an imaginative engineer in flight research, became fascinated by the M2, by now called the "Cadillac" for the two small fins emerging at the blunt tail. He built a successful flying model, which led to authorization for a manned glider.

In many ways the local authorization seemed more typical of the early NACA, since Headquarters did not know about it—nor did Langley, for that matter. But it appeared to be promising, and it could be done cheaply. One aircraft company later estimated that it would have cost at least $150,000 and considerable time to build the M2, but the Edwards crew did it quickly for less than $50,000. A nearby sailplane company built the laminated wooden shell (Reed was also an avid sailplane pilot), the landing gear was scrounged from a Cessna 150, and much of the other fabrication work was done by NASA personnel who were practiced hobbyists in the art of homebuilt aircraft. By 1963 the M2-F1, as it was now called, had been completed.

Initial flight tests required a ground vehicle to tow the M2-F1 above the dry lake bed, but none of NASA's trucks or vans possessed enough speed for the task. The inventive Edwards team decided to shop around for a hopped-up Pontiac convertible, further modified by a custom car shop in Long Beach to include roll bars, radio equipment, and special seats for observers. After results from the ground tow tests looked good, the next step involved aerial tow tests behind a C-47. By the time these flights concluded in 1964, the lifting body concept, despite its oddball history, seemed to be worth pursuing. NASA Headquarters and congressional people were both impressed. News reporters loved the lifting body saga, and there was keen interest in the more advanced lifting body designs that now became fashionable.

The M2-F1 showed the way, but far more work was needed involving high-speed descent and landing approach tests. By this time the air force became interested, and a joint lifting body program was formalized in 1965. Generally speaking, NASA, through the Flight Research Center at Edwards, held the responsibility for design, contracting, and instrumentation, while the air force

The North American XB-70 yielded valuable data on flight characteristics of large supersonic aircraft. In this view the no. 1 XB-70A (62-0001) is in a level cruise flight mode at a relatively high altitude, judging from the darkness of the sky.

supplied the launch aircraft for drop tests, assorted support aircraft, medical personnel, and the rocket power plant to be used in the advanced designs.

Northrop became the prime contractor for a group of aluminum "Heavy-weights" sponsored by NASA. At the time when arguments over a "dead-stick" shuttle reentry became hottest, some crucial HL-10 landing tests at Mach 1 speeds convinced planners that a shuttle without special landing engines could successfully complete reentry, approach, and landing. A final confirmation came during tests of the Martin X-24A (based on an air force project), whose shape was similar to a laundry iron. By the time the X-24A test flights ended (1969–71), designers had complete confidence in the ability of the proposed space shuttle to land on a conventional runway at the end of a space mission. The lifting body tests made an important contribution.

In other projects, explicit aeronautical research continued. At the Flight Research Center, another exotic plane captured the attention of flight aficionados—the Rockwell XB-70 Valkyrie, a Mach 3 high-altitude bomber. The air force began plans for the XB-70 in 1955, but by the time of its rollout cere-

monies in 1964, plans for a fleet of such large bombers had given way to reliance on advanced ICBMs with more powerful warheads. In the meantime,
the Kennedy administration had endorsed studies for a supersonic transport
(SST) for airline use, and the configuration of the XB-70 made it an excellent
candidate for flight tests in support of the SST program.

The XB-70 Valkyrie took to the air for the first time in the autumn of 1964.
With a fuselage length of 189 feet and a large delta wing measuring 105 feet
from tip to tip, its size, operating characteristics, and construction features
made it an excellent SST prototype. The air force and NASA began a cooperative test program with the XB-70 in the spring of 1966, the first airline-size aircraft in the world able to make sustained, long-range supersonic flights. The
flight requirements for a Mach 3 airliner similar to the XB-70 were far more
complicated than those for a Mach 2 aircraft, such as the Anglo-French Concorde SST. A Mach 3 airliner's structure required more exotic alloys, such as titanium, because the conventional aluminum airframe of a plane like the Concorde could not survive the aerodynamic heating at great speeds. Working with
the XB-70 uncovered a number of maintenance and operational issues. For example, integrating a large, Mach 3 aircraft into the nation's existing airway
traffic system uncovered a special problem, because it made turns that required
hundreds of miles to complete.

Despite the loss of one XB-70 in a midair collision, in which two test pilots
were killed, the NASA test program generated invaluable data on sustained supersonic flight. On one hand, XB-70 tests conclusively demonstrated that shock
waves from SST airliners would prohibit supersonic routes over the continen-
tal United States. These tests helped fuel the opposition to the American SST
program. On the other hand, the knowledge accumulated about handling qualities and structural dynamics supplied basic data for use in future supersonic
military aircraft and in high-speed airliners. But the test program proved to be
too expensive to sustain indefinitely. Early in 1969 the XB-70 Valkyrie made
its last flight, to the Air Force Museum in Dayton, Ohio.

When the political question arose as to whether the United States should
enter the international competition for a supersonic commercial transport aircraft—a sweepstakes already begun by Great Britain and France jointly with
their Concorde and by the Soviet Union with its TU-144—NASA already had
a solid data base to contribute. It also had the laboratories and the contracting
base to manage the program. But wise counsel from Deputy Administrator
Dryden led to NASA's retreat into a supportive R&D role; he argued that with
Apollo under way, NASA could not politically sponsor another high-technology,
enormously expensive program during the same budget years without one of

This haunting view of Earth rising above the desolate lunar horizon greeted the *Apollo 8* astronauts as they came from behind the Moon after the lunar orbit insertion burn. The lunar horizon is approximately 485 miles from the spacecraft. Earth is some 240,000 miles away.

them being sacrificed to the other or both of them killing each other off in competition for funds. The subsequent history of the SST program, including its eventual demise, seemed an eloquent testimonial to the wisdom of his judgment. He died in 1965; in 1976 Dryden Flight Research Center was named for him.

Apollo to the Moon

Although the tragic fire of January 1967 delayed plans for manned spaceflight in Apollo hardware for approximately eighteen months, the versatility of the system came to the rescue. Managers quickly shifted the burden of checking out the major components of the system to unmanned flights, while a quick-opening hatch was designed and tested, combustibles were sought out and replaced, and the wiring design was completely reworked. After a nine-month delay, flight tests resumed. On 9 November 1967, *Apollo 4* became the first unmanned launch of the awesome Saturn V. A 360-foot-high stack of three-stage launch vehicle and spacecraft, weighing 2,824 tons, slowly lifted off Launch Complex 39, propelled by a first-stage thrust of 7.5 million pounds. A record 278,000 pounds of payload and upper stage went into Earth orbit. Later the third stage fired to simulate the lunar trajectory, lifting the spacecraft combination to over 10,000 miles. With the third stage discarded, the service module fired its engine to raise the apogee to 11,000 miles, then burned again to propel the spacecraft toward Earth reentry at the 25,000-MPH return speed from the Moon. All systems performed well; the third stage could restart in the vacuum of space; the automated Launch Complex 39 functioned beautifully. This momentous mission convincingly vindicated the once-controversial concept of "all-up" testing.

Next came the unmanned flight of the laggard LM. On 22 January 1968 a Saturn 1B launched a 32,000-pound LM into Earth orbit. It separated, and it tested its ascent and descent engines. With this reassuring performance, the LM also passed its first flight test.

Procedures to man-rate the mammoth Saturn V followed in turn. *Apollo 6*, on 4 April 1968, successfully ran the launch vehicle through its paces—launch, separation of the stages, the guidance system, the electrical systems. This set the scene for the first manned spaceflight in Apollo since the tragic fire. NASA scheduled *Apollo 7* to test the crew and command module for the ten days in space that would later be needed to fly to the Moon, land, and return. But beyond *Apollo 7*, the schedule remained in real difficulty. It was the summer of 1968; only a year and a half remained of the decade within which this nation had committed itself to land astronauts on the Moon. Somehow the flight schedule had to be accelerated.

In Houston, George Low thought it could. After all, he reasoned, even the test-flight hardware was built to go to the Moon; why not use it that way? The advantages of early experience at lunar distances would be enormous. On 9 August he broached the idea to Gilruth, who enthusiastically endorsed it. Within days the senior managers of the program had been polled and had checked for problems that might inhibit a circumlunar flight. All problems proved to be fixable, assuming the *Apollo 7* went well. The trick then became to build enough flexibility into the *Apollo 8* mission so that it could go either way, Earth-orbital or lunar-orbital.

Apollo 7 soared aloft on 11 October 1968. A Saturn IB put three astronauts into Earth orbit, where they stayed for 11 days, testing particularly the command module environmental system, fuel cells, and communications. All came through with flying colors. On 12 November NASA announced that *Apollo 8* had been reconfigured to carry out a lunar orbit. It signified a bold jump.

Astronaut Neil A. Armstrong took this photograph of Buzz Aldrin deploying the passive seismic experiments at Tranquility Base, while the ungainly Lunar Module crouches in the background.

On 21 December a Saturn V lifted the manned *Apollo 8* off Launch Complex 39 at the Cape. Astronauts and ground crews confidently repeated the familiar phases: Earth orbit, circularizing the orbit, all as rehearsed. But then the Saturn third stage ignited again and added the speed necessary for the spacecraft to escape Earth's gravity on a trajectory to the Moon. On 23 December the three-man crew became the first human beings to pass out of Earth's gravitational control and into that of another body in the solar system. The TV camera looked back at Earth—a small, round, rapidly receding ball, warmly laced with a mix of blue oceans, brown continents, and white clouds that appeared with startling clarity against the blackness of space.

On Christmas Eve *Apollo 8* disappeared behind the Moon and out of radio communication with Earth. At that point, the crew became the first humans to see the mysterious back side of the Moon. Next, they fired the service module engine to reduce their speed enough to be captured into lunar orbit—irrevocably, unless the engine would restart later and boost them back toward Earth.

Another engine burn regularized their lunar orbit at seventy miles above the surface. Using on-board television cameras, the astronauts shared breathtaking views of the battered lunar landscape with millions on Earth. The crew read the biblical creation story from Genesis and wished viewers a happy holiday season. On Christmas Day they fired the service module engine once again, acquired the 3,280 feet per second additional speed needed to escape lunar gravity, and triumphantly headed back to Earth. They had at close range verified the lunar landing sites as feasible and proved out the hardware and communications at lunar distance, except for the all-important last link, the LM.

That last link still posed a major concern. NASA managers expended two more flights to confirm the LM's readiness for lunar landing. The *Apollo 9* flight (3–13 March 1969) became the first manned test of the LM during an Earth orbit mission. For assurance, NASA scheduled another test flight to be certain that everything worked after enduring a long voyage in the environment of space. On 18 May 1969 *Apollo 10* took off on a Saturn V to find out. The entire lunar landing combination blasted out to lunar distance. Once in lunar orbit, the crew separated the LM from the command module, descended to within nine miles of the surface, fired the ascent system, and docked with the command module. At last, all systems were "go."

On 16 July 1969 *Apollo 11* lifted off for the ultimate mission of Apollo. Saturn V performed beautifully, lofting the spacecraft combination toward the Moon. Once in lunar orbit, the crew checked out their precarious second home, the LM. On 20 July the LM separated and descended to the lunar surface. At 4:18 P.M. (EST) came the word from Astronaut Neil A. Armstrong: "Houston— Tranquility Base here—The *Eagle* has landed." After checkout, Armstrong set foot on the lunar surface: "that's one small step for [a]* man—one giant leap for mankind." The eight-year national commitment had been fulfilled; humans were on the Moon. Armstrong set up the TV camera and watched his fellow astronaut Edwin E. Aldrin Jr. join him on the lunar surface, as Michael Collins

*In the official NASA history of the lunar mission, *Chariots for Apollo: A History of Manned Lunar Spacecraft* (1979), the following explanatory note about the voice communications system (VOX) appears on p. 346: "Whether he actually uttered the article or not caused considerable discussion. Armstrong, himself, later wrote: 'I thought it had been included. Although it is technically possible that the VOX didn't pick it up and transmit it, my listening to the recording indicates it is more likely that it was omitted.'"

circled the Moon in the *Columbia* command module overhead. More than one-fifth of the Earth's population watched ghostly TV pictures of two space-suited men plodding around gingerly in an unlikely world of gray surface, boulders, and rounded hills in the background. The astronauts implanted the U.S. flag, deployed the scientific experiments to be left on the Moon, collected their rock samples, and clambered back into the LM. The next day they blasted off in the ascent module and rendezvoused with the command module.

The astronauts returned to an ecstatic reception. For a brief moment, millions of people around the world suspended their day-to-day divisions. Tuned to television and radio broadcasts, they took collective pride in Apollo's journey to the moon and pride in navigating this new ocean as humans continued their exploration of the cosmos.

Chapter 6

Aerospace Dividends, 1969–1973

The worldwide euphoria over NASA's stunning voyage to the Moon did not rescue the NASA budget. At NASA's moment of greatest triumph, Congress drastically cut back the space program from the $5 billion annual budgets that had characterized the mid-1960s. With the peak of Apollo production expenses now in the past, NASA expected part of the reduction. But the depth of the cut stemmed from emotional changes in the political climate, mostly centering on the unpopular Vietnam War—its escalating costs, the tragic loss of lives, and the divisive protests at home. As Congress read the public pulse, the cosmos could wait; the Soviet threat seemed less menacing; the new political reality lay in domestic problems. NASA watched as Congress chopped its fiscal 1970 budget to $3.7 billion. Something had to give. Although the agency continued its basic Apollo program, the last three flights had to be deleted. Space science projections took a hard hit. Ambitious programs for planetary exploration often dwindled into oblivion; those that survived did so as far more modest ventures. Some facilities simply disappeared from NASA's map. For example, the new Electronics Research Center in Cambridge, Massachusetts, under construction since 1964, wound up being transferred intact to the Department of Transportation—a $40 million facility taking with it 399 of 745 skilled employees. Nonetheless, even with diminished resources, NASA still conducted solid research and contributed significant results in the explorations of air and space.

Space Probes and Earth Satellites

But the bought and paid for projects continued to earn dividends. The Orbiting Astronomical Observatory (OAO-2) lifted off on 7 December 1968. It was the heaviest and most complex automated spacecraft yet in the space science program and took ultraviolet photographs of some twelve hundred objects for the first time—stars, comets, planets, and distant galaxies. OAO-2 delivered portentous results, including additional evidence of the existence of black holes in space. *Mariner 6* and *Mariner 7,* launched in early 1969, journeyed to Mars, flew past as close as nineteen hundred miles, and returned 198 high-quality TV photos of the planet.

Follow-on missions of OAO craft and Orbital Geophysical Laboratories turned up reams of fascinating new scientific information, and additional Mariner spacecraft successfully journeyed to Mars. Scientific results from all these ventures yielded riveting data about pulsars, titanic solar flares, Martian geography, and more. With the confirmation of black holes (the enigmatic collapsed star remnants so dense in mass and gravity that even light cannot escape) and the previous discoveries of quasars and pulsars, these findings added up to one of the most exciting decades in modern astronomy.

Planetary exploration opened further vistas of other worlds. *Pioneer 10,* launched 2 March 1972, left the vicinity of Earth at the highest velocity ever achieved by a spacecraft (32,000 MPH) and began an epic voyage to the huge, misty planet Jupiter. Steadily sailing past Jupiter and away from the Sun, in 1987 *Pioneer 10* crossed the orbit of Pluto, becoming the first manmade object to travel out of our solar system and into the limitless reaches of interstellar space.

Pioneer 10's partner, *Pioneer 11,* took off on 5 April 1973 to follow the same outward path. On 3 December 1974 it passed Jupiter at the perilously close distance of 26,000 miles—as opposed to 80,000 for *Pioneer 10*—and returned data. The composite picture from the reports of the two spacecraft depicted an enormous ball of hydrogen with no fixed surface, emitting much more radiation than it received from the Sun, shrouded with a turbulent atmosphere in which massive storms such as the Great Red Spot (25,000 miles in length) had raged for at least the four hundred years since Galileo first trained a telescope on Jupiter. *Pioneer 11* swung around the planet and, taking advantage of Jupiter's gravitational field, accelerated outward at 66,000 MPH toward the distant planet Saturn, where in 1979 it would observe at close range this lightest of the planets (it could float on water), its mysterious rings, and its 3,000-mile-diameter moon Titan.

Going in the other direction, *Mariner 10* left Earth on 3 November 1973, headed inward toward the Sun. In February 1974 it passed Venus, gathering information that confirmed the inhospitable character of that planet. Then, using Venus's gravitational force as propulsion, it charged on toward the innermost planet, Mercury. The cumulative evidence from the Mariner instruments pictured a planet essentially unchanged since its creation some 4.5 billion years ago, except for heavy bombardment by meteors, with an iron core similar to Earth's, a thin atmosphere composed mostly of helium, and a weak magnetic field.

In spite of the fascinating information about our fellow voyagers in the solar system, and as important as the long-range scientific consequences might be, Congress and many government agencies continued to be much more intrigued with the tangible, immediate-return, Earth-oriented program that began operations in 1972. On 23 July the Earth Resources Technology Satellite *(ERTS 1)* settled into a polar orbit. From that vantage point, it covered three-quarters of Earth's land surface every eighteen days, at the same time of day (and therefore with the same sun angle for photography), affording virtually global real-time information on developing events such as crop inventory and health, water storage, air and water pollution, forest fires and diseases, and recent urban population changes. In addition it depicted the broad area (and therefore undetectable by ground survey or aircraft reconnaissance) geologic patterns and coastal and oceanic movements. *ERTS 1* also interrogated hundreds of ground sensors monitoring air and water pollution, water temperature and currents, snow depth, and so forth and relayed information to central collection centers in nearly real time. The response was instantaneous and widespread. Foreign governments, states, local governments, universities, and a broad range of industrial concerns quickly took notice. They exploited data on excessive grazing of grasslands, violation of strip mining guidelines, identification of new fault lines in areas prone to earthquakes, and dozens of other issues useful to American citizens as well as the global community.

Like the experimental communications satellites of the early 1960s, the ERTS series found an immediate clientele of governmental and commercial customers clamoring for a continuing inflow of data. The pressure made itself felt in Congress; on 22 January 1975 *Landsat 2* (formerly *ERTS 2*) reached orbit ahead of schedule to ensure continuation of the data that *ERTS 1* (renamed *Landsat 1*) had provided for two and a half years, and a third satellite was programmed for launch in 1977. This schedule reassured experimental users of the new system that they could securely plan for continued information from the satellite system.

OAO-2, the orbiting astronomical observatory, was the largest, heaviest, and most complex scientific spacecraft NASA had developed at the time of its flight in 1969. With its solar panels deployed, as shown here, *OAO-2* was twenty-one feet long and weighed forty-four hundred pounds; it carried eleven ultraviolet telescopes into space.

The earth resources program had another important meaning. It constituted a quiet but dramatic shift in the nature and objectives of the space program. During the 1960s the Moon had been the big target, and large and expensive programs had been the name of the game. As the decade ended, it became increasingly clear to the NASA management that the political climate no longer supported that kind of space program. Legislators wanted to know how projects contributed to solving the everyday problems of their constituents. The Space Shuttle finally won approval with arguments for its heavy dedication to studies of our earth and its convincing economies in operation.

In another sign of the times, it seemed that NASA increasingly functioned as a service agency. In 1970 NASA for the first time launched more satellites for various customers (ComSatCorp, National Oceanographic and Atmospheric Administration, DoD, and foreign governments) than for itself. Five years before only 2 of 24 launches had been for others. For realistic planners in the agency, NASA's future appeared to be tied to this new trend.

Twilight for Apollo

Meanwhile, Apollo ran down its impressive course. *Apollo 12* (14–24 November 1969) repeated the *Apollo 11* adventure at another location on the Moon, the Ocean of Storms. Because *Surveyor 3* had been squatting there for two and a half years, it made an appealing target site. A pinpoint landing put the LM within six hundred feet—easy walking distance—of the Surveyor spacecraft. In addition to deploying scientific instruments and collecting rock samples from the immediate surroundings, Astronauts Charles (Pete) Conrad and Alan Bean cut off pieces from *Surveyor 3*, including the TV camera, for return to Earth and analysis after thirty months of exposure to the lunar environment.

Apollo 13 thundered into space on 11 April 1970 to continue lunar exploration. But fifty-six hours into the flight, well on the way to the Moon, the astronauts heard a disturbing "thump" in the service module behind them. An oxygen tank had ruptured. Pressure dropped alarmingly. As precious minutes ticked by, the astronauts and personnel in mission control could only speculate about the total damage, the effect on other systems, and the danger to the spacecraft combination. The backup analysis system on Earth sprang into action. Using the meager data available, the teams at contractor plants all over the country simulated, calculated, and reported. They delivered a grim verdict: *Apollo 13* appeared to be seriously, perhaps mortally, wounded.

Even on the shortest possible return path to Earth, the three men did not have enough air or water or electricity to sustain them. Ground crews and astronauts hurriedly assessed—and discarded—a list of emergency options. Suddenly, a solution seemed to present itself—why not retreat into the LM, a self-contained spacecraft unaffected by the disaster? The lunar landing was out of the question anyway; the lifesaving question was how to get three men around the Moon and back to Earth before their life-supporting consumables ran out. The simulations indicated that the LM could substitute for the command module, supplying propulsion and oxygen and water for an austere return trip. With emergency instructions from mission control, the *Apollo 13* crew reprogrammed their spacecraft to loop around the Moon and set an emergency course for home. The engine for the LM responded as hoped; off they went back to Earth. Still, the astronauts' descent through space remained a near thing—powered down to the point of minimum heating and communication, limiting activity to the least possible in order to save oxygen. Again, the flexibility and depth of the system came to the rescue. When conditions for reentry appeared to be within the limited capabilities of the crippled Apollo, the crew jettisoned the "lifeboat" LM along with the wounded service module. *Apollo 13* reentered safely.

After delays to complete appropriate spacecraft changes, *Apollo 14* lifted off on 31 January 1971, its mission focused on the scientific exploration of the moon. *Apollo 15* introduced the "Moon car," the Lunar Rover. With this electric-powered, four-wheel-drive vehicle, developed at Marshall at a cost of $60 million, the astronauts roamed beyond the narrow confines of their landing site. *Apollo 16* arrived in the Descartes area in April 1972, stayed 71 hours, and acquired photos and measurements of lunar properties. *Apollo 17,* launched 7 December 1972, ended the Apollo program with the most productive scientific mission of the lunar exploration program. The site, Taurus-Littrow, encompassed the oldest and the youngest rocks on the lunar surface. For the first time a trained geologist, Harrison H. Schmitt, was on a crew, adding his professional observations. EVA time totaled over 22 hours, and the Lunar Rover traveled some twenty-two miles.

With these flourishes, the Apollo program concluded. From beginning to end, it lasted eleven and a half years, cost $23.5 billion, landed twelve men on the Moon, and produced an overwhelming amount of evidence and knowledge. Technologically, it generated hardware systems several orders of magnitude more capable than their predecessors. In various combinations, the components of this technology could be used for a wider variety of explorations. Scientific analysis would continue for decades. The Lunar Receiving Laboratory, constructed in Houston, became the "archive" of the 840 pounds of physical lunar samples returned from various parts of the Moon. Scientists in this country and fifty-four foreign countries went to work, analyzing samples with a broad range of instruments and the expertise of many scientific disciplines. Gross results had already established that the Moon was a separate entity from Earth, formed at the same time as Earth some 4.5 billion years ago. Apollo bequeathed social legacies as well. For example, the program demonstrated the ability of a diversified system of government, industry, and universities to mobilize behind a common national purpose and produce on schedule an immense and diverse system directed to a common purpose.

Anticipating future manned missions using the Space Shuttle, NASA found ways to keep its mission specialists sharp, launch crews up to par, and contractors ready. Using leftover hardware from Apollo/Saturn stocks, NASA's next chapter in manned spaceflight featured the Skylab. Trimmed back to one orbital workshop and three astronaut flights, Skylab endured a hectic financial and planning career, the converse of Apollo. The revised plan called for an

Jupiter, its Great Red Spot, and three of its four largest satellites are visible in this photo taken 5 February 1979 by *Voyager 1*. The spacecraft was 17.5 million miles from the planet at the time. The innermost large satellite, Io, can be seen against Jupiter's disk. This photo was assembled from three black and white negatives by the Image Processing Lab at Jet Propulsion Laboratory.

S-IVB stage of the Saturn V to be outfitted as a two-story orbiting laboratory, one floor for living quarters and the other for working space. As the major objective, Skylab's managers wanted to determine whether humans could physically withstand extended stays in space and continue to do useful work. The most ambitious hardware development involved the Apollo telescope mount. Its five major experiments covered the entire range of solar physics and made it the most powerful astronomical observatory ever put in orbit. Other major areas of investigation called for earth resources observations and medical experiments involving the three-man crew. A roster of important subcategories of activities included an electric furnace to explore possibilities of using the weightless environment to perform certain industrial processes. Given the environment of gravity on Earth, forming perfectly round ball bearings or growing large crystals, much in demand in the electronics industry, remained elusive. But in the near-weightless conditions of space, Skylab held the promise of a manufacturing breakthrough.

On 14 May 1973 a giant Saturn V lifted off from Kennedy Space Center to place the unmanned 165,000-pound orbital workshop in Earth orbit. Within minutes after launch, disquieting news filtered through the telemetry reports from the Saturn V. The large, delicate meteoroid shield on the outside of the workshop had apparently been torn off by the vibrations of launch, causing serious damage to the two wings of solar cells designed to supply most of the electric power to the workshop. Once the workshop reached orbit, the news worsened. The loss of the big shade exposed the metal skin of the workshop to the hot sunshine; internal temperatures soared to 126°F. Astronauts attempting to board Skylab would be threatened not only by the heat but also by poisonous gases released by equipment on board.

With the launch of the first crew on hold, a far-flung support team worked around the clock for ten frantic days, trying to improvise fixes that might salvage the $2.6 billion program. With only partial knowledge of the precise degree and nature of the damage, engineers had to work out fixes that met the known problems yet retained versatility to cope with unknown ones. Two major efforts evolved: one to devise a deployable shade that the astronauts could spread over the metal surface of the workshop, the other to assemble a versatile tool kit of cutters and snips to release the solar wing from whatever prevented it from unfolding.

On 25 May 1973, a Saturn 1B lifted an Apollo command and service module combination into orbit, and the module docked with the workshop the same day. The crew entered the workshop the next day. In the course of two demanding and risky EVAs, they erected a makeshift shade and freed an effective section of the solar wing. The remaining Skylab missions were almost anticlimactic after the dramatic rescue of the workshop. With only minor problems, the missions ticked off their complicated schedules of experiments. After 28 days, the first crew undocked and returned to Earth. The second crew, launched 28 July 1973, spent nearly 60 days in orbit; the third crew, launched 16 November 1973, completed 84 days in orbit.

Scientists reckoned that the vast mass of astronomical and earth resources data from the Skylab program would keep investigators busy for years to come. The medical data and the industrial experiments yielded more immediate results. With the corrective exercises available on Skylab, there seemed to be no physiological barrier to the length of time humans could survive and function in space. Biological functions stabilized after several weeks in zero gravity. The industrial experiments on the melting and solidification process in weightless conditions looked promising; single crystals grew five times as large as those producible on Earth. Some high-cost industrial processes apparently possessed new potential if conducted outside Earth's gravitational effects.

As the empty Skylab continued to circle the Earth, its orbit began to decay, threatening an uncontrolled reentry. NASA regained some control over the rogue Skylab in the spring of 1979 and managed to steer it to reentry over the Indian Ocean. Still, chunks of the Skylab made a fiery plunge into remote areas of Australia, a reminder of the potential dangers of civilization's own debris from the exploration of space.

Transonic and Hypersonic Flight Research

Although questions about an SST aircraft persisted, NASA and its principal contractor, Boeing, kept working on the design throughout the 1960s. By 1971 production plans were under way when the program came to a halt. Critics remained adamant about the costs of the SST and its ability to operate economically. Flight tests of the big XB-70 Valkyrie had done little to quell the issue of sonic booms, and worrisome questions about adverse environmental effects at high altitudes continued. Congress finally voted against funds for construction of an SST for flight testing.

The British and the French proceeded with a smaller SST, the jointly developed Concorde, which began flight tests in 1969 and entered service in 1976. A Soviet SST, the Tupolev TU-144, also began internal schedules in 1976, but it was withdrawn from service two years later. Meanwhile, NASA and American aerospace companies cooperated in a research effort known as the Supersonic Cruise Aircraft Research Program. Beginning in 1973, this activity involved analysis of propulsion systems and advanced airframes. The ongoing SST studies, which continued into the 1980s, made considerable progress in quieter, cleaner engines as well as much improved passenger capacity and operational efficiencies. If the opportunity for second-generation SST airliners materialized later, NASA and the aerospace industry intended to lead the way with an American design.

While investigation of the supersonic regime continued, a major breakthrough at the transonic level occurred—the supercritical wing. The transonic regime had beguiled aerodynamicists for years. At transonic speeds, both sub-

David R. Scott, commander of *Apollo 15*, works behind the lunar rover at the Hadley-Apennine landing site. Hadley Rille is at the right center of the picture. Hadley Delta, in the background, rises approximately 13,124 feet above the plain.

sonic and supersonic flow patterns encased an aircraft. As the flow patterns went supersonic, shock waves flitted across the wings, resulting in a sharp rise in drag. Since most commercial jet airliners operated in the transonic range, coping with this drag factor could bring major improvements in cruise performance and yield substantial benefits in operating costs.

During the 1960s Richard Whitcomb committed himself to a program intended to resolve the transonic problem. For several years Whitcomb intensely analyzed what came to be called the "supercritical" Mach number—the point where the airflow over the wing went supersonic, with a resultant decline in drag. Analysis and wind tunnel tests led to a wing with a flattened top surface (to reduce its tendency to generate shock waves) and a downward curve at the

An overhead view of the Skylab Orbital Workshop in Earth orbit as photographed from the *Skylab 4* command and service modules (CSM) during the final fly-around by the CSM before returning home. The gold "parasol," clearly visible in the photo, was designed to replace the missing meteoroid shield, which also protected the workshop against solar heating.

trailing edge (to help restore lift lost from the flattened top). But wind tunnel tests were one thing. Real planes in the air were often something else. The next step required thorough flight testing of a plane equipped with the unusual wing.

Given its reduced budgets, NASA continued to make a virtue of improvising. Fortunately, engineers came up with an available plane that lent itself to comparatively easy modification: the Vought F-8A Crusader. The structure of the plane's shoulder-mounted wing made it easy to remove and replace with the supercritical design. Moreover, the landing gear of the F-8A retracted into the fuselage, leaving the experimental wing with no outstanding production encumbrances. The navy had spare planes available, and its speed of Mach 1.7 made it ideal for transonic flight tests. Although the test plane had begun life as a navy fighter, NASA aimed the supercritical wing program primarily at civil applications. The airlines as well as the airline manufacturers closely followed the development of the new airfoil.

The modified Crusader, designated the TF-8A, made its first flight at Edwards in 1971 and continued for the next two years. The test flights yielded data that corresponded to measurements from the preliminary tunnel tests at Langley. Most important, the supercritical wing promised genuine improvement in the transonic region, a fact that translated directly into reduced fuel costs and lower operational costs. Ironically, foreign manufacturers of business jets became the first to apply the new technology in new designs like the Canadair Challenger (Canada) and the Dassault Falcon (France). At the same time, both Boeing and Douglas applied the concept in experimental air force transports

NASA selected a Vought F-8A Crusader as the test-bed aircraft (designated TF-8A) to install an experimental supercritical wing in place of the conventional wing. The unique design of the supercritical wing (SCW) reduces the effect of shock waves on the upper surface near Mach 1, which in turn reduces drag.

like the YC-14 and the YC-15. In short order the supercritical wing became virtually standard on airliners and similar aircraft—American as well as foreign—and on certain military designs.

As additional commercial manufacturers began utilizing data from the supercritical wing studies, NASA and the air force collaborated in the development of its military applications for combat planes. Known as TACT, for Transonic Aircraft Technology, the military effort used a modified F-111A. By the early 1980s, with refined flight testing of the F-111A still continuing, several operational aircraft had been designed to utilize information from this project.

NASA's use of military aircraft to probe the transonic region paralleled a different effort that involved very high supersonic speeds. The aircraft in this case was one of the most exotic creations to fly—the Lockheed YF-12A, a highly classified interceptor design that led to the equally highly classified SR-71A Blackbird reconnaissance aircraft. According to published performance figures, the Blackbirds were capable of Mach 3 speeds at altitudes of eighty thousand feet or more. The planes originated in the famed Lockheed "Skunk Works" of Clarence "Kelly" Johnson, where Johnson and a talented group of about two hundred engineers put aeronautical pipe dreams on paper and then proceeded to build and fly them. The operating requirements of the plane at extreme speeds and altitudes for sustained periods created a completely new regime of requirements for parts and systems. As Johnson commented later, virtually every component of the aircraft, from its propulsion system and airframe materials to its rivets, fasteners, and hydraulic fluids, had to be developed as a new invention.

The first Blackbird flew in 1962; NASA first became involved in 1967, when Ames, where early wind tunnel data was acquired under tight security, was given permission to use the data in ongoing research. In return the Flight Research Center at Edwards organized a small team to assist the air force flight tests. But NASA wanted its own Blackbird for tests that would support the SST program still under way in the late 1960s. By this time the SR-71A was operational, and the air force had put two YF-12A prototypes in storage at Edwards. When the air force offered the pair to NASA, the agency quickly accepted and also assumed operational expenses, although the air force assigned a small team for assistance in maintenance and logistics.

NASA launched its Blackbird program with great enthusiasm. Engineers from Lewis, Langley, and Ames had a keen interest in propulsion research, aerodynamics, structural design, and the accuracy of wind tunnel predictions involving Mach 3 aircraft. The first YF-12A test missions under NASA jurisdiction began late in 1969, and flights averaged once a week during the next ten years; an impressive variety of high-speed problems were examined. One series involved a biomedical team who monitored physiological changes in the flight crews in order to measure stress in the demanding environment of high-speed operations. Many Blackbird test flights routinely carried instruments to analyze boundary layer flow, skin friction, heat transfer, and pressures in flight. Various structural techniques were employed in test panels on the planes. An experimental computerized checkout system diagnosed problems in flight and provided information for maintenance that would be required before the next mission. The checkout system was seen as a valuable asset for application in the Space Shuttle as well as military and commercial planes.

In many ways the Blackbird program, covering a decade of intensive flight tests, exemplified one of the Flight Research Center's most useful programs, providing a rich legacy of information for later aircraft built for sustained cruise at Mach 3. The end of the program prompted a chorus of protest from the Blackbird flight team and other NASA personnel who felt that the United States was frittering away its lead in high-speed flight and in technology generally.

Such grumbling proved to be premature. The American interest in aerospace and the national commitment to new technology remained strong, although it took different directions. At first glance, the new concern for controlling aircraft noise, reducing pollutants from engines, and enhancing overall aircraft fuel efficiency might have seemed less glamorous than derring-do at Mach 3. But the rationale for confronting such issues became urgent in the late 1970s, and the solutions to these issues were no less complex and challenging than the problems of high-speed flight. Aeronautical research continued to be a dynamic field of NASA programs to come. At the same time, in comparison to the United States, European progress in the fields of aeronautics and astronautics continued to gain momentum.

Chapter 7

International Ventures, 1973–1980

As work on an advanced space transportation system—the Space Shuttle—progressed, NASA negotiated a historic agreement to join American and Soviet spacecraft in Earth orbit. Mars became the target of instrumented space probes, and several international programs achieved striking success, including notable excursions into remote solar regions. NASA's aeronautics research paid off in the form of significantly improved aircraft engines for both large and small aircraft. Using a skillfully adapted Boeing 737 airliner, another program advanced operational procedures and safety in the increasingly complex air traffic control networks. Also, a varied program of aeronautical testing explored various modes of low-speed flight and a unique design for tilt-wing aircraft.

New Partners in Space

In the process of fabricating components for Skylab, NASA moved ahead with significant plans for future ventures in space. The initiatives bore the imprint of Thomas O. Paine, acting administrator after Webb's resignation in 1968 and administrator of NASA from March 1969 until he returned to industry in September 1970. One goal involved a broad approach to increased cooperation in space exploration. Like so many of America's international space initiatives in the postwar period, this effort offered separate proposals to the Soviet Union and to western European countries.

Beginning in 1968, the approach to the Soviet Union included suggestions for advanced cooperation, especially in the expensive arena of manned spaceflight. Because the Soviets had always evinced a strong concern for the safety of their cosmonauts, American planners suggested a joint system for international space rescue. Prospects seemed good, owing to the changed dynamics of the space race. For both the United States and the Soviets, absolute secrecy surrounding space missions no longer loomed as a major issue, because modern tracking systems yielded enough details to determine the size of payloads, booster characteristics, and destinations. In any case, by 1969 the evidence was clear that whether or not the Soviet Union had in fact been in a Moon landing race with the United States, the United States had taken the lead.

As a first step, Paine broached the idea of a Soviet linkup with the Skylab orbital workshop, but the idea of USSR crews arriving at an American space outpost as guests had little appeal for the Soviets. Further discussions turned up lively Soviet interest in a completely new project to develop compatible docking and rescue systems for manned spaceflight. Negotiations proceeded rapidly. Completed by George M. Low, acting administrator after Paine's departure, the grand plan for the Apollo-Soyuz Test Project (ASTP) called for a mutual docking and crew exchange mission that could develop the necessary equipment for international rescue and establish international rescue criteria for future manned systems from both nations. The proposal called for an Apollo spacecraft to rendezvous and dock with a Soyuz already in orbit. Using a specially developed docking unit between them, they would adjust pressurization differences of the two spacecraft and spend two days docked together, exchanging crews and conducting experiments. Both sides rapidly agreed to engineering guidelines. In the process the ASTP also became a political test for the validity of the detente agreements that President Richard M. Nixon had negotiated with the Soviet Union.

An unprecedented process ensued, requiring detailed cooperation between the two superpowers. A series of joint working groups of Soviet and American specialists met over several years to work out the various hardware details and operational procedures. At the Nixon-Brezhnev summit in 1973, discussions on the ASTP initiative narrowed the prospective launch date to July 1975. The most concrete example of U.S.-USSR cooperation in space proceeded with good faith on both sides. The mission flew as scheduled on 15 July and smoothly fulfilled all objectives.

The Space Shuttle

The other major initiative from Paine began on the domestic front and then expanded to the international arena. President Nixon appointed a Space Task Group to recommend broad outlines for the next ten years of space exploration. Within this group, Paine won acceptance for the concept of the Space Shuttle. In its original conception, the Space Shuttle resembled a rocket-boosted, airplanelike vehicle designed to take off from a regular airport runway and climb up to orbital speed and altitude. Once in orbit, crews might deploy satellites, repair or retrieve satellites already in orbit, and, using an additional Space Tug stage, transport manned and unmanned payloads throughout the solar system. The launcher and the shuttle would be reusable for up to one hundred flights, halving the cost per pound in orbit. Also, the design of the satellites could focus on orbital rigors, since the satellites would not endure the additional stress of rocket launch.

Paine left NASA to return to industry; his successor, James C. Fletcher, took office in April 1971 and immediately reviewed the status of the Space Shuttle, particularly for its political salability. He quickly became convinced that the shuttle, at an estimated cost of $10.5 billion, was too expensive to win congressional approval. Fletcher instigated a rigorous restudy and redesign, which cut the cost in half, mainly by dropping the plan for unassisted takeoff and substituting two external recoverable, reusable solid rockets and an expendable external fuel tank. The revised version of the Space Shuttle finally gained the necessary congressional support, and President Nixon approved its development in January 1972.

First Paine and then Fletcher tried to get a commitment from western European nations to supply major systems for the shuttle. Their own joint space program had not been an unqualified success. In 1964 western European nations had organized two international space organizations, European Launch Development Organization (ELDO) to produce launch vehicles and European Space Research Organization (ESRO) to produce spacecraft and collect and interpret results. Despite Europe's technical capability, issues of assigning specific contracts to separate countries and allocating budgets hampered European progress. A proposed booster featured three stages, each built by a different country. The launch record reflected a gloomy history of one kind of failure after another. After years of effort, western Europe had little to show for its independent launch vehicle. Much had been learned about multinational coordination of advanced technology, however, and successful joint projects like Concorde supersonic transport and several multinational military aircraft ven-

The Space Shuttle *Atlantis* atop the shuttle carrier aircraft returns to the Kennedy Space Center after a ten-month refurbishment in 1998.

tures (such as the Panavia Tornado) had promoted a sophisticated aerospace community in Europe. Moreover, using American boosters, the ESRO group had successfully launched a variety of scientific satellites, applications satellites, and space probes. In addition to experienced contractors, the European space organizations had also developed international centers, for example, the European Space Research and Technology Center (ESTEC) in the Netherlands, to carry out research and manage space projects. By the early 1970s, a consensus in Europe endorsed the need for a new, unified organization. Based on the growing capabilities of its aerospace community, the European Space Agency (ESA) began operations in 1975. It was a new start.

Into this restive environment, Paine arrived to talk about the next generation of the U.S. space program. He extended the promise of some discrete major segment to be developed and produced in Europe—a partnership that would give the European nations a meaningful piece of the action with full pride of useful participation. The Europeans warmly responded, though it took a while for a project to coalesce. Finally, a joint agreement emerged: western Europe agreed to build the self-contained Spacelab that would fit in the cargo bay of the shuttle spacecraft. This pressurized module would provide a shirt-sleeve environment in which scientists could conduct large-scale experiments. In addition, an unpressurized scientific instrument pallet would give large telescopes and other instruments direct access to the space environment. The estimated cost for these elements came to $370 million. The partnership with Europe expanded in 1975 when Canada joined the international effort, agreeing to foot the $30 million R&D bill for the remote manipulator system used to emplace and retrieve satellites in orbit.

The Space Shuttle era opened completely new prospects for spaceflight: extensive international collaboration; nonpilots in space; multiple payloads to be delivered as required or picked up out of orbit; and new designs of satellites, free from the expensive safeguards against the vibrations and shocks of launch by rocket.

During the late 1970s, the Space Shuttle became the largest consumer of NASA's budget and of the attention of its managers. Since its beginnings in the early 1970s, the development story for the Space Shuttle differed considerably from that of Apollo in the 1960s. The projected costs had been halved to win the necessary political approval of the program. NASA managed to achieve this cut to $5 billion by making severe compromises in the original design—from a system that would take off from a runway like an airplane, fly into orbit, and return to land on a runway like an airplane, to a system that would take off vertically like a rocket, jettison the boosters and fuel tanks, and return to land on a runway like an airplane. Other compromises followed, as the budget continued to be lean year after year. Trying to maintain some semblance of a schedule, NASA and its contractors sometimes set aside potential development problems until money became available to investigate them. By 1977 a modified Boeing 747 carried a prototype shuttle aloft, where it detached and carried out approach and landing flights at Dryden Flight Research Center. Progress seemed consistent, if occasionally irregular. But postponed technological issues had a way of becoming insistent problems.

In 1978 it became obvious that serious questions continued to dog the main engines. Aided by two solid-rocket boosters, the orbiter mainly relied on a cluster of three of these high-pressure liquid-hydrogen-fueled engines to propel itself into orbit. Not only were the main engines expected to produce the highest specific impulse of any rocket engine yet flown, but they also had to be throttleable and reusable—they had to fire again and again for many flights before being replaced. Stubborn shortcomings persisted. By 1979 a series of painstaking, component-by-component analyses had identified and fixed most of the problems. Individual engines recorded better test runs—until the first firings of the clustered engines generated a new set of difficulties. Grudgingly, they finally yielded to concentrated engineering rework.

The other pacing item on the orbiter concerned the thermal protection tiling to shield most of the orbiter surface from the searing heat of reentry. Manufacture and application of the thirty-three thousand tiles lagged too far behind, disrupting other crucial schedules. Early in 1979 NASA decided to ferry the orbiter from the manufacturer's plant in California to Kennedy Space Center so that the remainder of the tiles could be applied there while technicians completed other work and system checks. But problems still festered. The tiles were brittle and easily damaged; they did not bond to the metal properly, and thousands had to be reapplied. Between the tiles and the engines, the Space Shuttle ran over its budget for several years, and the date for the first flight slipped two painful years behind. All this meant serious consequences for many government, domestic, and international customers.

By the end of 1980, however, prospects improved for the first flight to occur in the spring of 1981. Meanwhile, the other three orbiters moved at a steady pace through manufacturing. Encouraged, NASA posted a schedule for operational flights that left the Space Shuttle booked solidly out to the middle of the 1980s.

The Planets

In space science the big program continued to revolve around Viking, the first major initiative in a decision NASA had made some years before: to focus the space science program on the planets. Apollo, the reasoning went, would keep scientists busy for years analyzing the mass of data and samples gathered from the lunar surface. Not until that information had been assimilated would there be a need to consider whether investigators needed more information from the Moon and, if so, what kind.

Meanwhile, space science, though not neglecting the study of the Sun and the universe, would concentrate on the inner planets of our solar system and begin an assault on the enigmatic outer planets. Along with results from Apollo, the early planetary flights confirmed that every celestial body had worthwhile lessons to teach—scientific lessons, important in their own right, as well as lessons that illuminated problems on earth. Comparative planetary studies could shed light on the nature of Earth's minerals and their proportions. The perplexities of tectonic plates and volcanism in the universe continued to raise many questions. How to explain oceans and the unique atmosphere of Earth in comparison with neighboring planets? Why does our atmosphere circulate and transfer heat the way it does? Every new body we studied represented a new laboratory and a different set of data. Following Apollo, the planetary ventures of the 1970s often received less press attention, but they introduced dramatic new vistas and provided insights into our solar system.

Mars, the most likely of the inner planets, became the first target of an ambitious planetary venture. In two launches the Viking program proposed to deploy four spacecraft in the vicinity of Mars. The spacecraft arrived in the vicinity of Mars in mid-1976 and achieved successful orbits. From that vantage point, the scheme called for two orbiters to photograph the surface and serve

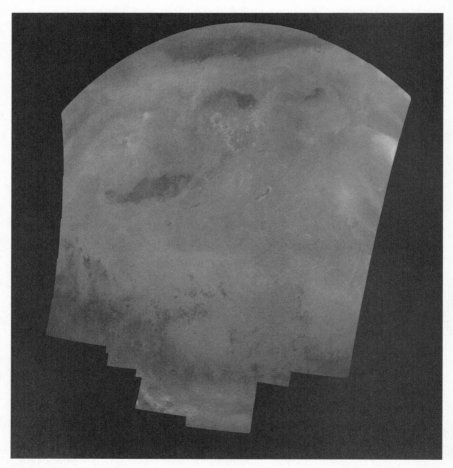

This mosaic is composed of 104 Viking Orbiter images acquired on 11 February 1980. At that time it was early northern summer on Mars. A major geologic boundary extends across this mosaic, with the lower third of the image showing ancient cratered highlands; north of the boundary are the lowland northern plains.

Voyager I obtained this dramatic view of Jupiter's Great Red Spot and its surroundings on 25 February 1979, when the spacecraft was 5.7 million miles from Jupiter. Cloud details as small as 100 miles across can be seen here.

as communications relays. About the same time, two landers headed for the rocky Martian surface to photograph the terrain, measure and monitor the atmosphere and climate, and conduct chemical and biological tests on the soil for evidence of rudimentary life forms. All of this amounted to a daunting performance, requiring very ambitious technology and complex science to be operated from a distance of more than 40 million miles. But perform Viking did, in a technological triumph comparable to the Apollo landings on the Moon.

The orbiters' depiction of the Martian surface, plus the weather and seismic reports from the landers, yielded a profile of a static planet with evidence of a geologically tumultuous past. The surface of Mars included towering volcanoes half again as high as any on Earth along with vast, eroded canyons deeper than any on Earth. The evidence pointed to intense volcanic action about 3 billion years ago. Scientists detected water, frozen in broad polar ice caps; seasonal and diurnal variations in temperature; dust storms; traces of seismic activity; and a thin atmosphere with trace gases that suggested a denser atmosphere sometime in the past. Despite all these clues of a fascinatingly diverse history, Viking's instrumentation uncovered no indications of life within the hostile environment of Mars.

Scientists also targeted Earth's nearest planetary neighbor, Venus, for probes during the last half of the 1970s. Two Pioneer spacecraft departed toward Venus in the summer of 1978. Studying Venus presented a notably different problem than studying Mars or Earth. Its thick, heavy, hot atmosphere was impervious to normal photography and could be depicted only by means of radar. The first spacecraft, arriving at Venus in December 1978, carried mapping radar to delineate the major features on the surface. Mission officials described the second spacecraft as a "bus" that released four probes in a broad pattern. The probes parachuted slowly through the atmosphere, sending back measurements until they crashed. The Venusian atmosphere, they reported, had a high sulfur content, with oxygen and water vapor at lower levels. By 1980 the orbiter had mapped more than 80 percent of the Venusian surface, which resembled a vast, rolling plain punctuated by features resembling two continents and a massive chain of islandlike sites.

Study of the outer planets using more sophisticated spacecraft began in 1977 with the launch of *Voyager 1* and *Voyager 2* on eighteen-month flights to Jupiter. The Voyager system, *Science* magazine reported, contained technology that was 150,000 times more advanced than that of the *Mariner 4* system, which flew to Mars in 1965. *Voyager 1* made its closest approach to Jupiter in March 1979, and *Voyager 2* followed in July. The sensors recorded in fine-grain detail the intricate weather patterns on Jupiter and detected massive lightning bolts in the cloud tops. Passes by the planet's nearby moons turned up striking evidence of active volcanoes on one of them, Io.

With a boost from Jupiter's gravitational field, the Voyagers set their course for distant, ringed Saturn, where *Voyager 1* arrived in November 1980 and *Voyager 2* arrived in August 1981. With sufficient control gas remaining, the mission extended to a far-away Uranus flyby in January 1986; a Neptune flyby was planned for August 1989.

In the study of the Sun and its interrelationships with Earth, several missions embodied extensive coordination with European space organizations. Launched in 1976, *Helios 2* was part of a joint program with the Federal Republic of Germany to study basic solar processes. As the solar cycle moved toward its maximum phase, NASA dispatched the Solar Maximum Mission in

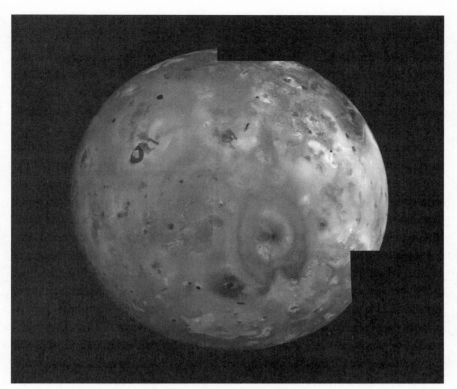

Perhaps the most spectacular of all the Voyager photos of Io was this mosaic obtained by *Voyager 1* on 5 March 1979 at a range of about 248,000 miles. After observing a great variety of color and albedo on the surface, scientists concluded that they were probably the result of surface deposits of various forms of sulfur and sulfur dioxide. The two great volcanoes Pele and Loki *(upper left)* are prominent.

1980 to study solar flares, although problems with the satellite led to a rendezvous and retrieval by a shuttle crew in 1984. To study the effects of solar radiation on Earth's magnetosphere and atmosphere, NASA launched *International Sun-Earth Explorer 1* and *2* (*ISEE 1* and *2*) in 1977. Positioned some distance apart but in similar elliptical orbits, the two satellites (one provided by NASA, the other by the ESA) monitored the complex interactions of Earth's magnetosphere with incoming solar radiation. In 1978 the international partners added *ISEE 3* to the system. In a wider study of the universe, the major program of the second half of the 1970s consisted of a series of three highenergy astronomy observatories. Carrying instruments from NASA, the United Kingdom, and the ESA, the *International Ultraviolet Explorer (IUE)* recorded ultraviolet emissions using two ground control centers from which experimenters could direct the observations of the satellite. These various ventures exemplified a growing trend of international collaboration in both manned and unmanned space exploration. Also, they allowed all the participants—including NASA—to stretch their respective budgets. In the years just ahead, such trends proved instructive for NASA and congressional committees when debating the costs and technological developments required to get a manned space station into orbit.

Closer to home, in the latter half of the 1970s, NASA heeded a congressional mandate to organize an intensive study of Earth's upper atmosphere. In this activity both NASA and national legislators responded to growing environmental concern about the effects of gases such as Freon on the ozone layer. A continuous measuring program resulted; several agencies provided data from which a detailed model of the complex processes could be constructed. Reflecting a continuing interest in the management of earth resources, *Landsat 3* was launched in 1978 to maintain the flow of data to the growing number of users. An ambitious new earth resources program targeted potential clients in the agricultural community in the United States as well as abroad.

With the launch of *Seasat 1,* NASA attempted a new form of surveying maritime conditions and resources. Intended to report on such variables as sea temperature, wave heights, surface-wind speeds and direction, sea ice, and storms, *Seasat 1* achieved instant success. Unfortunately, its life was cut short after three months in orbit by electrical power failure. Nonetheless, enough data had been recorded to verify the effectiveness of the instrumentation and the existence of a group of potential users in the weather, maritime, and fisheries communities. Toward the end of the 1970s, NASA also made contributions to environmental research with different satellites for weather observation and geophysics. These led to enhanced interpretation of data in weather forecasting, mineral surveys, and earthquake studies.

Aircraft and the Environment

In keeping with the rising energy concerns of the 1970s, NASA committed considerable resources to new engine and aircraft technologies to increase flight efficiency as a means of conserving fuel. The Aircraft Energy Efficiency Program, inaugurated in 1975, developed fuel-saving techniques for application in current aircraft as well as future designs. The project covered several areas of investigation: more efficient wings and propellers, composite materials that were lighter and more economical than metal, improved fuel efficiency in jet engines, and new engine technologies for aircraft in the future.

The supercritical wing—a research initiative that promised rich dividends—represented only one aspect of activity that also led NASA into the arcane subject of laminar flow control. A smooth flow of air over the surface of a plane, or laminar flow, is a characteristic of low speeds. At cruising speeds the airflow becomes turbulent, creating increased drag. A carefully articulated research program yielded deeper understanding of this phenomenon that affected many types of commercial and military aircraft. Advanced research in the field relied on some unusual hardware development. Using models and analytical testing, NASA developed a system of tiny holes on the wing surface and a lightweight suction system to draw off the turbulent air. By the late 1980s, the agency prepared to begin flight testing of a laminar flow control system for possible use on commercial aircraft.

Other research efforts were carried out through the Engine Component Improvement Program. The objective was to target engine components for which wear and deterioration led directly to decreased fuel efficiency in jet engines. As a result of this research, new components to resist erosion and warping were introduced, along with improved seals, ceramic coatings to improve performance of gas-turbine blades, and improved compressor design. Research results were so positive and so rapidly adaptable that new airliners of the early 1980s, such as the Boeing 767 and McDonnell Douglas MD-80 series, used engines that incorporated many such innovations.

For business jets, NASA rebuilt an experimental turbofan, incorporating newly engineered components designed to reduce noise. Completed by 1980, this project successfully developed engines that generated 50 to 60 percent less noise than current models. For larger transports, Lewis Research Center started tests of two research engines that cut noise levels by 60 to 75 percent and reduced emissions of carbon monoxide and unburned hydrocarbons as well.

In a different context, NASA became engaged in procedures for flight operations in congested air space. Among the issues that needed assessment were aircraft noise during landing and takeoff over populated areas, safe approach and landing procedures in bad weather, and methods for controlling high-density traffic patterns. NASA's Langley Research Center became the lead organization for this research and took charge of test protocols. Extraordinarily useful information emerged from studies using a modified Boeing 737 twin-jet transport. In the plane's passenger area, NASA technicians put together a second cockpit equipped with the latest innovations in instrumentation. This second cockpit became the flight center for research operations; the crew occupying the standard cockpit in the 737's nose functioned as a backup. It carried the programmatic label of Terminal Configured Vehicle/Advanced Transport

Operating System (TCV/ATOPS). Beginning with flights in 1974, the plane experienced a productive career before its retirement in 1993. In addition to testing new equipment for precision descent and approach procedures on instruments, the plane played a key role in demonstrating the Microwave Landing System in 1979. The International Civil Aviation Organization eventually adopted the Microwave Landing System, over a competing European design, as the standard system to be used around the world.

Moreover, Langley's 737 performed yeoman service in demonstrating a wide range of cockpit technologies, including satellite-based Global Positioning System equipment and the design of "glass cockpits" using computerized displays to depict the status of aircraft systems, navigational information, and other data. This seemingly mundane kind of R&D actually carried considerable weight in advancing the state of the art for cockpit design and its acceptance by airline executives and the crews who piloted the aircraft. Langley's research tasks with the 737 took place in an increasingly problematic technological environment, where airline executives and manufacturers of airline transports seemed to be swayed more by real-world demonstrations than by research papers published in an academic journal. Coming aboard and flying with the 737 team, then debriefing themselves in postflight discussions with TCV/ATOPS personnel often carried the day in winning acceptance for new and unfamiliar operational procedures and equipment. The eventual installation of proven technologies enhanced the economical performance of airline equipment and vastly improved the safety of airline travelers.

In another series of dramatic excursions involving the 737, the Federal Aviation Agency also became a partner in a program to deal with deadly microbursts, a weather phenomenon in stormy conditions that generates abrupt and powerful downdrafts near the ground. For aircraft in the process of reducing airspeed and descending for final landing approach, microbursts could—and did—trigger tragic crashes. Although ground-based equipment sometimes helped detect these meteorological killers, it was essential to develop some sort of aircraft-based detection system. Langley specialists collaborated with the weather service to locate suitably dangerous locales, dispatching the NASA plane to fly directly into harm's way. Concluding a challenging R&D program that progressed into the early 1990s, the redoubtable 737 and its research teams led the way in concocting a system that provided crucial seconds of warning about microbursts ahead of the aircraft. Manufacturers, leaning on NASA's invaluable data, quickly produced the requisite instrumentation for airliners in the United States and around the globe.

At Ames, scientists became interested in using aircraft as platforms for in-

The XV-15 tilt-rotor research aircraft at liftoff. For cruising, the engines and rotors pivot forward to a horizontal position to give cruising speeds twice those of helicopters. The XV-15s, manufactured by Bell, were involved in limited research at NASA/Dryden in 1980 and 1981. Development of the XV-15 tilt-rotor research aircraft began in 1973 with joint army-NASA funding as a "proof of concept," or "technology demonstrator," program, with two aircraft being built by Bell Helicopter Textron in 1977.

vestigations of terrestrial as well as astronomical phenomena. Beginning in 1969, Ames acquired a number of different research planes and launched several imaginative investigations that continued over the following decades. High-altitude missions relied on a pair of Lockheed U-2 aircraft, originally supplied to the air force as reconnaissance planes. They carried out earth resources observations, compiled land usage maps, surveyed insect-infested crops, and measured damage from floods and forest fires. The high-flying U-2 aircraft provided information covering hundreds of square miles; for a more intensive look at details in a smaller area, Ames brought in other specialized planes that flew midaltitude missions.

One of the pioneers in midaltitude missions was a refurbished airliner—a Convair 990 christened the *Galileo*. The four-engine jet began operating in the early 1970s and conducted a variety of tasks, such as infrared photography, detection of forest fires, and meteorological investigations. Over the Bering Sea in 1973, a joint study with the Soviet Union gathered data on meteorological phenomena, ice flow, and wildlife migratory patterns. The first Convair was lost in a tragic midair collision with a navy patrol plane, but its operations had been so productive that acquisition of a second plane was authorized, and *Galileo II* went to work in 1974. Conducting research at midaltitude heights, the new Convair 990 carried out international missions as well, including archaeological studies of Mayan ruins and observations of monsoon patterns in the Indian Ocean.

Other planes were added, including the small Learjet and the huge Lockheed C-141 Starlifter, which became operational with the Ames fleet in 1974. The Starlifter's interior size and load-carrying capacity made it the best candidate for installation of a 915-centimeter telescope for astronomical observations. Many of the C-141 missions, as well as those involving other Ames research planes, were international in scope. In 1977 the C-141, known as the Kuiper Airborne Observatory, flew to Australia to make observations of the planet Uranus during especially favorable astronomical conditions. American and Australian scientists studied the planet's atmosphere, composition, shape, and size, and discovered that Uranus possessed equatorial rings.

At about the same time, the Learjet, equipped with a 30-centimeter infrared telescope, was operating high over the Arctic on a different international mission, known as Project Porcupine. Ames worked with the Max Planck Institute of Physics and Astrophysics in a study of the coupling between the magnetosphere and the ionosphere. The experiment called for the launch of a sounding rocket from Sweden. After the rocket ejected a barium charge, the Learjet followed the barium trail along Earth's magnetic lines of force. Collectively, these researches by aircraft on a global scale enhanced professional contacts for NASA personnel and generated favorable foreign press coverage for the agency as well as for the United States.

As Ames proceeded to carve out its niche in using aircraft as research platforms, the center also strengthened its role in flight research and moved beyond wind tunnel testing to flight testing. Taking advantage of congressional support for aeronautical research, the director of Ames, Hans Mark (appointed 1969), guided the center into research on short-haul aircraft, including vertical/short takeoff landing (V/STOL) designs. Since the mid-1960s Ames had been working with the U.S. Army on helicopter research, relying on the big low-speed tunnels at Ames, along with its excellent simulator equipment and other facilities. By the 1970s both the FAA and the air force were working with Ames on a new generation of short-takeoff transports. In 1976, to the chagrin of Langley, Ames officially became NASA's lead center in helicopter research.

Although the Pioneer project and future planetary missions shifted to JPL at the same time (completed by 1980), the new aircraft programs enlivened activities at Ames.

Among the rotor-craft investigations, one of the most interesting involved the XV-15 Tilt Rotor Research Aircraft, with wingtip-mounted engines. For takeoff and landing, the engines remained vertical, with the big rotors providing lift; once in the air, the engines and rotors tilted to the horizontal, propelling the XV-15 forward. This design offered the flexibility of a conventional helicopter, able to land and depart from confined areas but also capable of faster speed and longer range. Bell Helicopter Textron built two aircraft for NASA and the army. The first XV-15 went to Ames in 1978 for extensive tests in the 40-by-80-foot wind tunnel, to be followed by flight tests at Bell's plant in Texas. Test pilots undertook the first demonstration of in-flight tests of the two prototypes at Ames and at Dryden Flight Research Center during 1980.

Somewhat more conventional in design, the Quiet Short-Haul Research Aircraft evolved from a NASA initiative to investigate new technologies for commercial airliners. The research plane emerged as a hybrid, based on an extensively modified de Havilland C-8A Buffalo. Under contract to NASA, Boeing rebuilt the plane with new avionics, new wings and tail, and a quartet of jet engines mounted above the wing to generate "upper surface blowing" in order to increase lift. The plane made its maiden flight at Boeing's Seattle plant in 1978, then flew to Ames for continued flight tests. The short takeoffs and quiet operation of the aircraft yielded considerable information for application in both civil and military design. One intriguing series of tests led to a successful landing and takeoff from an aircraft carrier—the first four-engine jet plane to accomplish this feat.

For NASA the decade of the 1980s seemed particularly satisfying. Its aeronautical programs turned out to be successful and productive, space science recorded solid achievements, linking with the Soviet Union in space helped reduce cold war tensions, and progress in the Space Shuttle raised everyone's confidence for prospects of outstanding missions to come. That confidence was to be severely tested.

Chapter 8

Aircraft and Aerospace Craft, 1980–1989

For NASA flight research, the 1980s opened with a significant administrative change—the Dryden Flight Research Center lost its independent status and became a directorate of Ames Research Center in 1981. Nevertheless, several exotic flight programs emerged during the decade, and a variety of unusual aircraft continued to populate the skies above Edwards Air Force Base. Spaceflight continued to occupy much of NASA's time and budget. Bureaucratic changes in this arena included the reorganization of the National Space Technology Laboratories (formerly the Mississippi Test Facility) in 1988 as the John C. Stennis Space Center. The new Stennis facility not only became the focus of rocket and propulsion testing but also developed additional expertise in remote sensing, an area of increasing activity for NASA. The catastrophic loss of a shuttle and its crew, plus a string of other failures, shook NASA's management; resolute managerial changes and encouraging success in other programs eventually helped to maintain sorely tested morale.

Diversity in Flight

Given the cost of experimental flight aircraft and the evolution of increasingly sophisticated electronic and simulator systems, it was perhaps inevitable that engineers would turn to smaller, pilotless radio-controlled aircraft. In the 1980s this idea was embodied in the HiMAT (the full name was Highly Maneuverable Aircraft Technology). Powered by a General Electric J85 turbojet engine, the HiMat had a length of 23 feet and a wingspan of 16 feet.

The compact HiMAT embodied an evolutionary concept, originating during the M2 lifting body program of the 1960s. To test a variety of lifting body shapes in flight, an innovative NASA engineer at Edwards built a twin-engine radio-controlled model that carried the smaller test models high into the sky and made 120 test drops. Typical remotely piloted vehicles (or RPVs) used an autopilot system and had restricted maneuverability. The Edwards aircraft, in contrast, was completely controlled from the ground, using instrument references. By the late 1960s, Edwards personnel were flying an actual lifting body test configuration, the Hyper III, in drop tests from a helicopter. Veteran fliers who flew the model by remote control found it a remarkable experience. "I have never come out of a simulator emotionally and physically tired as is often the case after a test flight in a research aircraft," one pilot said. "I was emotionally and physically tired after a 3-minute flight of the Hyper III," he admitted. Although remote flight research continued, demands of the YF-12 Blackbird program and other projects kept it at a low level. Still, significant progress occurred.

The team at Dryden took a Piper Twin Comanche, fitted with an electronic fly-by-wire system, added a television system for a remote pilot, and turned it into a successful remotely piloted aircraft from takeoff to landing. Although a backup pilot flew in the cockpit, the remote operators practiced stalls, stall recoveries, and even made precise instrument landing approaches. In the early 1970s, these skills were translated into an applicable test program to investigate stall and spin phenomena after several fighter planes were lost in spinning accidents. NASA let contracts to McDonnell Douglas for three three-eighths-scale models of the F-15. Each model cost $250,000; a full-sized plane cost $6.8 million. Piloted from the ground and released from a B-52 at high altitude, the model F-15 program yielded valuable information, which was applied in final revisions of the operational air force fighter. The remote pilots doing the flying found the spin tests quite challenging: the heartbeats of pilots went from the 70–80 per minute of normal, manned flights to 130–40 during the remotely piloted drop tests.

The remotely controlled flight tests generated brisk controversies in various coteries of the flight test community. It was not just that a steely-eyed test pilot no longer settled into the cockpit and sallied forth to execute a series of gut-wrenching maneuvers beyond the ken of mere mortals back on the ground. Extensive ground support systems proved to be nearly as expensive for remote flight operations as for manned aircraft. But remotely controlled flights were useful; models offered a cost-effective method for testing esoteric designs; they possessed obvious advantageous in dangerous flight maneuvers. The positive factors became increasingly convincing as NASA and the military services pondered aerial demonstration of exotic configurations and materials for combat planes in the 1990s and beyond. The logic for a test vehicle like the HiMAT could not be dismissed.

The HiMAT structure itself evolved as a melding of ingredients composed of various metal alloys, graphite composites, and glass fiber materials. It had sharply swept wings, winglets, and canard surfaces—considered aeronautically avant-garde when the first plane flew in 1978. Carried aloft by a B-52, controllers flew the HiMAT remotely—and safely—through a series of complex maneuvers at transonic speeds. Engineers designed the HiMAT as a modular vehicle so that wings, control surfaces, and structural materials could be removed, replaced, and evaluated at a fraction of the cost of building a full-sized aircraft. The HiMAT's changing configurations suggested the possible shapes of aircraft to come.

While the HiMAT continued to test alternative design ideas, flight test specialists nonetheless recognized the persistent value of full-sized manned aircraft. Consequently, NASA decided to proceed with the Grumman X-29, a plane whose dramatic configuration matched that of the HiMAT. The X-29 had a single vertical tail fin and canard surfaces—not necessarily unique in the 1980s. What made the X-29 so fascinating was its sharply forward-swept wings.

The forward-swept wing had precursors in German designs of World War II. In 1944 Junkers put such an experimental jet into the air—the JU-287. The war ended before extensive flight tests could be carried out, but the JU-287 quickly revealed one of the major problems of any swept-forward design: structural divergence. Lift forces on wings cause them to bend slightly upward. When the wings sweep forward, this force tends to twist the leading edge upward, increasing lift and the bending motion until the wing fails. One solution involved structural design to keep the wing absolutely rigid, but conventional metal construction made such wings so heavy that they became impractical. Although swept-forward wings occasionally appeared on various aircraft in the postwar era, construction and weight problems proved intractable. The solution appeared in the form of composites, which could be used to fabricate wings of light weight but high strength.

Grumman had submitted an earlier, unsuccessful HiMAT design, which ran into severe wing-root drag problems. A forward-swept wing seemed to offer answers, and the company had quietly pursued the idea. NASA also became interested, and the DoD eventually agreed to support a radical new design. NASA became responsible for technical support and flight testing. In 1987 the plane was officially introduced as the X-29, the first new "X" aircraft developed by the United States in more than a decade. The fuselage took shape very quickly, since the forward section came from a Northrop F-5A. Landing gear came from the General Dynamics F-16A, and the engine evolved from a General Electric power plant developed for the McDonnell Douglas F-18 Hornet. At first glance, the X-29 seemed a sorry aeronautical compromise, merely incorporating bits and pieces from other planes. But its wings and related design elements made it truly unique. Moreover, it was highly unstable, requiring a trio of flight computers to keep it under control while responding to the pilot's inputs.

When the X-29 made its first flight in 1984, the forward-swept wings and canard surfaces were its most distinguishing characteristics. In aircraft with swept-back wings, controllability became a problem as increasingly turbulent air flowed over the wingtips and tail surfaces. The X-29's wingtips, however, were always moving in comparatively undisturbed air, enhancing controllability at high speeds, and the canard surfaces also operated in an airstream much less turbulent than that around the tail. The rigid wing of the X-29 owed much to composites and the way they were layered in relation to the angle of the wing and aerodynamic stresses, overcoming the tendency to structural divergence.

Exotic experimental military planes were only one of several areas of NASA's study. During the 1970s the general aviation sector became increasingly robust. Most Americans knew little about this remarkably diverse segment of American aviation, which included all aircraft except those flown by commercial airlines and the armed services. There were about 2,400 scheduled airliners in service during the 1970s and 4,300 in the 1980s, while the general aviation fleet climbed from 150,000 to 220,000 aircraft, ranging from propeller-driven single engine planes to multimillion-dollar executive jets. Sales of general aviation aircraft contributed significantly to America's favorable balance of payments, since 90 percent of the world's fleet of general aviation types originated in American factories. Given the scope of general aviation operations in the United States and the significance of American domination of the world market for this sector, NASA's attention was probably overdue when the agency began comprehensive studies during the late 1970s. Results came very quickly as more than a dozen production and prototype designs incorporated features derived from relevant NASA studies.

One distinctive hallmark of NASA's general aviation investigations featured the wingtip winglet, a device to smooth out distorted airflow and thereby improve wing efficiency and enhance fuel economy. During the 1980s a number of high performance business jets, such as the Learjet, as well as late-model transports built by Boeing and McDonnell Douglas, used this innovation. The agency also developed a new high performance airfoil for general aviation, the GAW-1. A separate research effort went into stall/spin problems, using radio-controlled scale models as well as several different full-sized operational aircraft. NASA mounted additional programs to probe exhaust and engine noise, engine efficiency, and the use of composites. A special investigation of crash survivability tested the airframes of planes and injuries to passengers, represented by carefully instrumented anthropomorphic dummies. A huge drop tower let the test planes plunge onto a typical runway; the test results proved useful to many aviation industry firms, including structural engineers and manufacturers of aircraft seats, seat belts, and body restraint systems.

Satellites and Space Science

During the 1970s the number of American payloads put into space by rocket boosters diminished as mission planners waited for the shuttles to become operational. When the shuttles began flying with payloads in the 1980s, this did not mean that NASA's expendable rocket launches ceased. Several rocket launches had already been scheduled, and NASA also intended to maintain this capability as a backup through the mid-1980s. NASA boosters orbited a variety of communications and environmental satellites as well as several spacecraft involving space science. Moreover, the audacious Voyager continued its richly rewarding "grand tour" of the outer planets. Shuttle launches may have received the lion's share of news coverage, but rocketed payloads continued to demonstrate their share of utility and value in space exploration.

Meteorological satellites and other Earth-oriented spacecraft expanded their essential roles in contemporary society. During 1981 another in a series of Geostationary Operational Environmental Satellites *(GOES-5)* went into Earth-synchronous orbit. In addition to expanded hurricane observations in the Caribbean zone, *GOES-5* tracked Gulf Stream currents for fishing crews and others with marine interests, provided invaluable data for weathercasters, and warned citrus growers about potentially crop-killing frosts. The National Oceanic and Atmospheric Administration (NOAA) not only supplied vital data on ocean temperatures and wave patterns with *NOAA-7;* the multimission spacecraft conducted a variety of atmospheric and tidal measurements while it monitored solar particle radiation in space, alerting manned space missions and commercial aircraft to potentially hazardous conditions.

This network expanded with the launches of *GOES-6* and *NOAA-8* in 1983.

The HiMAT (Highly Maneuverable Aircraft Technology) subscale research vehicle, seen here during a research flight, was flown by NASA/Dryden from mid-1979 to January 1983. The aircraft demonstrated advanced fighter technologies that have been used in the development of many modern high-performance military aircraft.

The latter joined a space-based search and rescue system cooperatively operated by the United States, France, Canada, and the Soviet Union and known as the Sarsat-Cospas network. The satellites of the participating countries could pinpoint the locations of emergency beacons aboard ships and aircraft in distress. Within a few months after it became operational, the rescue network had saved some sixty lives around the globe. *Landsat-4,* launched in 1982, experienced transmission failures, so *Landsat-5* took over during 1984, continuing vital coverage for forestry, agriculture, mineral resources, and other uses. Also during the 1980s, NASA launched a series of new Intelsat communications satellites to replace older models in geosynchronous orbits above the Indian, Pacific, and Atlantic Oceans. In myriad ways, spacecraft and space technology continued to demonstrate their value and gain global support.

Nonetheless, space science payloads and planetary probes continued to be the most dramatic performers. Following the encounter of *Voyager 1* with Saturn in 1980, *Voyager 2* made an even closer pass in the summer of 1981. These visits turned up much new information on Saturn's rings, moons, and weather systems and posed a number of new questions for planetary scientists. Continuing analysis of Pioneer *Venus 1* also seemed to raise as many new issues as it resolved. Launched in 1983, the Infrared Astronomical Satellite was a joint project of NASA and scientific centers in the Netherlands and Great Britain. During its ten-month lifetime, the international satellite detected new comets, analyzed infrared signals from a number of new galaxies, and yielded data that suggested that many of them may be merging or colliding with each other.

Planetary probes continued to turn up surprising insights into the nature of our solar system. Four and a half years after uncovering a wealth of new data on Saturn and its spectacular rings, *Voyager 2* approached Uranus in January 1986. By the time the intrepid Voyager completed its flyby, the spacecraft had revealed more information about the planet and its company of moons than observers had learned since its discovery by the English astronomer William Herschel more than two hundred years ago.

The spacecraft's arrival achieved something of a tour de force for the Jet Propulsion Laboratory (JPL), managers of Voyager's aptly named "Grand Tour of the Solar System." JPL's navigators had to place the spacecraft less than two hundred miles from a point between the planet's innermost moon, Miranda, and the planet's rings. Having traveled 1.8 billion miles from Earth, *Voyager 2* now whipped toward its goal at fifty times the speed of a pistol bullet. Commands from JPL to Voyager took two hours and forty-five minutes to arrive. Unless the JPL crew did everything correctly, *Voyager 2* might miss the gravitational sling from Uranus to send it on toward its rendezvous with Neptune in 1989. More important, engineers had to know the exact location of the Voyager so that its cameras would record something planetary instead of the infinite blackness of space. "That feat," explained a reporter for the *National Geographic*, "is equivalent to William Tell shooting an arrow in Los Angeles and hitting an apple in Manhattan." In order to avoid many potential glitches, controllers improvised new procedures, such as breaking into the onboard computer programs to fine-tune the thrusters; commandeering another backup computer to improve the rate of image processing; and dispatching further signals to help Voyager perform in a colder, darker environment than it had for its Saturn flyby.

There were other snarls as well, but *Voyager 2* carried on superbly, providing evidence of ten new moons besides the five known orbs circling Uranus. Miranda, the smallest of the five, proved especially dramatic with a tortured surface that included an escarpment ten times deeper than the Grand Canyon. The various moons constituted a geological showcase, with mountains up to twelve miles high, plains dotted with craters, and sinuous valleys that may have been gouged out by glaciers. *Voyager 2* also captured other curiosities about Uranus, including its offset magnetic field; its fascinating ultraviolet sheen, called an "electroglow"; and its erratic atmospheric patterns. Another mission to Uranus might be decades or even centuries away. But the Voyager's legacy promised to give scientists, space physicists, and astronomers considerable data to ponder in the meantime.

Shuttle Operations

At liftoff, the shuttle looked and sounded like an oversized rocket booster with wings. The power for the launch came from a combination of propulsion systems. A pair of solid-fuel booster rockets straddled a huge propellant tank filled with liquid hydrogen and liquid oxygen; the shuttle itself perched atop the cylindrical walls of the propellant tank, which fed the trio of Space Shuttle main engines mounted in the shuttle's tail. During the initial ascent phase, all five propulsion systems drove the shuttle upward. Following burn-out of the solid-fuel boosters, the empty casings separated from the external tank and parachuted back to earth, where recovery teams retrieved them from the ocean, refurbished them, and packed them again with segments of solid fuel. The shuttle's liquid hydrogen main engines continued to fire, drawing propellants from the external tank. When the tank was empty, it was finally jettisoned and destroyed by intense heat during its descent through Earth's atmosphere. A pair of maneuvering engines plus batteries of small rocket thrusters on the orbiter refined its orbital path as needed and provided maneuvering capability during the mission.

Compared to the Apollo spacecraft, the orbiter was huge, with a length of 120 feet and a wingspan of 80 feet. During a typical mission, seven crew members could live and work in the flight deck area, and the cargo bay contributed an additional payload or workspace area measuring 60 feet long by 15 feet in diameter. The designers sized the shuttle to carry payloads of 65,000 pounds, to orbit at an altitude of 230 miles (smaller payloads allowed orbits of up to 690 miles), return to Earth, and land with payloads of 32,000 pounds (such as a malfunctioning satellite). NASA contended that the ability to reuse the booster rocket casings and the ability of orbiters to make repeated missions made the Space Shuttle an extremely cost-effective space vehicle for years to come. Because of all the tiles on the orbiter, the personnel associated with the program often joked about the "flying brickyard," but there was great enthusiasm about the Space Transportation System, or STS.

Although launches occurred at the Kennedy Space Center (KSC), and the plans called for most orbiter flights to finish there on a special landing strip three miles long, there were alternative landing sites at Vandenberg Air Force Base and Edwards Air Force Base in California, at White Sands, New Mexico, and at selected emergency runways around the world. The first few landings were planned for the broad expanses of the dry lake at Edwards; the orbiter would be carried back to KSC from any remote site atop the specially modified Boeing 747 ferry aircraft. There were only five landings at Kennedy Space Center before a blown nose wheel tire at the end of the sixteenth mission shifted more touchdowns to Edwards. Some earlier flights had been diverted from Kennedy because of weather; the Boeing 747 transporter definitely proved its value in returning orbiters from Edwards, White Sands, and Vandenberg. Following the nose wheel incident, engineers planned changes for orbiter landing gear as well as improvements to the Kennedy landing site.

Concerns about tiles and engines kept *Columbia,* the first orbiter for flight missions, grounded at KSC for nearly two years. In the meantime, other shuttle crews kept their flying skills sharp by participating in further drop tests of the *Enterprise* and by training flights in a Grumman Gulfstream modified to imitate an orbiter's landing characteristics. Crew members and trainees practiced experiments and other tasks in a microgravity environment through long training missions in a converted Boeing C-135 transport. These missions also tested theories about the nature of nausea (motion sickness) caused by disorientation in space—a severe problem for crew members during long space missions. The plane would fly high, arching parabolas in the sky, giving trainees several seconds of "weightlessness" at the top of each stomach-churning climb. The training missions might last several hours—repeated climbs, nose-overs, and rapid descents before the next upward surge. For those aboard the plane, all of this could be either exhilarating or loathsome. Officially, NASA's C-135 was designated the Reduced Gravity Aircraft; unofficially, hapless trainees dubbed it the "weightless wonder," "vomit comet," and "barf buzzard."

Finally, long hours of flight training and grueling sessions in electronic simulators came to an end. The *Columbia's* flight crew, astronauts John Young and Robert Crippen, joked that they had spent so much additional time in the electronic simulators that they were "130 percent trained and ready to go." Their inaugural flight was set for 10 April 1981. But the *Columbia* mission, like others to follow, was scrubbed at the last minute on a technicality. Two days later, the countdown for *Columbia* matched a day of perfect weather at KSC, and the Space Shuttle thundered off into space, boosted by 7 million pounds of thrust from its solid-fuel rockets and liquid-hydrogen engines.

The No. 1 X-29 advanced technology demonstrator aircraft banks over desert terrain near NASA/Dryden. It flew in a joint NASA–air force–Defense Advanced Research Projects Agency program from December 1984 to 1988, investigating handling qualities, performance, and systems integration on the unique forward-swept-wing research aircraft. Phase 2 of the X-29 program involved aircraft No. 2 and studied the high-angle-of-attack characteristics and military utility of the X-29.

Reaching an altitude of 130 nautical miles, the *Columbia*'s crew settled into orbit for a two-day mission. The orbiter carried no cargo except an instrumentation package to record stresses during launch, flight, and landing, plus a variety of cameras. One of these, a remote television camera aboard the orbiter, revealed gaps around the tail section, where some tiles had apparently worked loose during launch. As the crew prepared for descent back to Earth, the mission controllers became quietly concerned, worried that other tiles in critical areas along the orbiter's underside might have fallen off as well. At a blinding speed of Mach 24, *Columbia* began its searing reentry back into Earth's upper atmosphere, where the intense heat of atmospheric friction built to over 3,000°F. Some anxious moments ensued as the plummeting spacecraft became enveloped by a blanket of ionized gases that disrupted radio communications.

At 188,000 feet, as the *Columbia* slowed to Mach 10, mission control heard a welcome report from Crippen and Young that the orbiter was performing as planned. A long, swooping descent and a series of planned maneuvers bled off the excess speed and brought the spacecraft in over the Edwards area. Parked in cars, jeeps, and campers all around the edge of the landing area, an estimated five hundred thousand people gathered to observe the shuttle's return. The sharp crack of a sonic boom snapped across the desert, and the crowd soon saw the *Columbia,* now slowed to about 300 MPH, make its final descent and touchdown, a true "spaceliner," marking a new era in astronautical ventures.

For all its teething problems, the shuttle performed remarkably well through five years and twenty-four successful missions. Inevitably, there was some fine-tuning and reworking of numerous tiles before a second launch of *Columbia* in November, the first time a spacecraft had returned to orbit. During 1982 three more missions marked the end of flight tests and the beginning of missions to deploy satellites. The next year, four additional missions included three in the new orbiter, *Challenger,* and ended with *Columbia*'s flight with the ESA's Spacelab aboard. There were six crew members, a record number for a single spacecraft, including Ulf Merbold, a German who represented the ESA. These flights in 1983, which carried America's first woman in space (Sally Ride) as well as the first black American (Guion Bluford), not only launched additional American and international payloads but also significantly increased activities in space science, particularly with the Spacelab mission. To deploy satellites from the cargo bay, the crew relied on a unit called the Propulsion Assist Module, or PAM, which was introduced on the STS-5 mission in 1982. In the payload deployment sequence, the remote manipulator system lifted the satellite out of the orbiter cargo bay. The orbiter then maneuvered away; the PAM attached to the satellite automatically fired about forty-five minutes later, boosting the payload higher. An independent system on the satellite then took over, using thrusters on the satellite to circularize its orbit, checking out its systems, and making the satellite operational. Although the PAM booster was augmented by other systems, many payloads could be left in orbit after simply lifting them out of the cargo bay with the remote manipulator system.

The orbiter *Discovery* joined the fleet in 1984, and *Atlantis* followed in 1985. The demographics of the orbiter crews reflected growing diversity, including more women; Canadians; Hispanics; Asians; assorted Europeans; a Saudi prince; a senator, E. J. "Jake" Garn; and a congressman, Bill Nelson. The various missions engaged astronauts in extended extravehicular activity, such as untethered excursions using manned maneuvering units. In Mission STS-11 (41-C) in 1984, an astronaut using one of these units assisted in the first capture of a disabled satellite, the Solar Maximum payload *(Solar Max)*; it was then repaired and redeployed. The mission also had the task of placing a new satellite in orbit. Scheduled for deployment was the Long Duration Exposure Facility, a twelve-sided polyhedron measuring 14 feet in diameter and 30 feet long. It carried several dozen removable trays to accommodate fifty-seven experiments put together by some two hundred researchers from eight countries. After being lifted out of *Challenger,* the big structure was to stay in orbit for a year, awaiting its return on a later shuttle flight.

For the crew aboard *Challenger,* the biggest task involved the first planned repair of a spacecraft in orbit. *Challenger'*s thrusters boosted it three hundred miles higher to intercept the *Solar Max* satellite. After some difficulties, owing to the satellite's tumbling motion, it was finally stabilized and cranked down into the cargo bay by the remote manipulator system (RMS). After a night's rest, George Nelson and James van Hoften donned space suits and went to work on the balky satellite; they replaced a faulty attitude control module and some electronic equipment for one of its instruments. The *Solar Max* was sent back into orbit, its repair job in space saving millions of dollars. Later the same year, during STS-14, the crew of *Discovery* had to retrieve a pair of errant satellites that had been placed in improper orbits by faulty thrusters. Although the RMS managed to capture the satellites, they would not drop into the cradles in the cargo bay for their return to Earth, and the mission specialists had to manhandle each one aboard before closing the cargo bay doors. These missions conclusively demonstrated the shuttle's ability to recover, repair, and if necessary, refuel satellites in orbit. The DoD also made two classified missions in 1985.

Mission STS-22, in October 1985, was the fourth Spacelab flight and was notable for its eight-member crew—the eighth person had to sleep aboard the Spacelab itself. Most significant was the special role of the West German Federal Aerospace Research Establishment, which managed the orbital workshop in which the Spacelab mission specialists carried out experiments in materials processing, communications, and microgravity. It was a highly successful mission, with only one memorable drawback. Aboard the Spacelab was a new holding pen for animals that contained two dozen rats and a pair of squirrel monkeys. The crew soon complained to controllers that the animal quarters needed modifications for any future flights. The food bars for the rats began to crumble, so that loose particles of rat food began floating around the Spacelab. Worse, some waste products from the rats also began to litter the Spacelab's atmosphere, leading to pointed scatological comments from the disgruntled crew.

Continuing missions carried a variety of American and international scientific experiments. One involved electrophoresis, in which an electric charge

Winglets proved effective in the design of the Learjet Model 55, shown here in flight during a NASA test program in 1984.

was used to separate biological materials; the goal in this case was to produce a medical hormone. Other experiments had to do with such topics as vapor crystal growth, containerless processing, metallurgy, atmospheric physics, and space medicine. The payload manifests for most missions became recognizably similar, listing satellites, experimental biomedical units, physics equipment, and so on. The manifest for STS-16 in 1985 possessed a decidedly different quality, including a pair of satellites along with a "Snoopy" top, a wind-up car, magnetic marbles, a pop-over mouse named Rat Stuff, and several other toys, including a yo-yo. For die-hard yo-yo buffs, a NASA brochure reported that the "flight model is a yellow Duncan Imperial." The news media gave considerable attention to the whimsical nature of the Toys in Space Mission, although the purpose was educational. As crew members videotaped the toy experiments, the astronauts demonstrated each toy and provided a brief narrative of scientific principles, including different behaviors in the space environment. The taped demonstrations became a favorite with educators—and the astronauts obviously delighted in this uncustomary mission assignment.

Tragedy

Despite occasional problems, shuttle flights had apparently become routine—an assumption that dramatically changed with *Challenger*'s mission on 28 January 1986.

On the morning of the flight, a cold front had moved through Florida, and the launch pad glistened with ice. It was still quite chilly when the crew settled into the shuttle just after 8:00 A.M. Many news reports remarked on the crew's diversity; seven Americans who seemed to personify the nation's heterogeneous mix of gender, race, ethnicity, and age. The media focused most of its attention on Christa McAuliffe, who taught social studies at a high school in New Hampshire. She was aboard not only as a teacher but also as an "ordinary citizen," since Space Shuttle missions had seemed to become so dependable. Scheduled for a seven-day flight, *Challenger* also carried a pair of satellites to be released in orbit.

NASA officials, leery of the icy state of the shuttle and the launch pad, waited two extra hours before giving permission for launch. When the Shuttle's three main engines ignited at 11:38 A.M., the temperature still hovered at about 36 degrees, the coldest ever for a shuttle liftoff. After a few seconds, the solid-fuel boosters also ignited, and *Challenger* thundered majestically upward. Everything appeared to be working well for about seventy-three seconds after liftoff. At forty-six thousand feet in a clear blue sky, the shuttle became virtually invisible to the exhilarated spectators at Cape Canaveral. But the telephoto equipment of television cameras captured every moment of the fiery explosion that destroyed *Challenger* and snuffed out the lives of its crew. In the aftermath of the tragedy, stunned government and contractor personnel took action to recover remnants of the shuttle and to begin a painstaking search for answers.

Answers were essential, because the three remaining shuttles were grounded while the cause of the *Challenger* explosion was identified and corrected. Until that time the United States could not put astronauts into space or launch any of the numerous satellites and military payloads designed only for deployment from the shuttle cargo bay. Moreover, construction of the planned space station in Earth orbit relied entirely on the shuttle's cargo capacity.

Detailed analysis of photography and shuttle telemetry pointed to a joint on the right solid booster. It appeared that a spurt of flame from the joint (which joined fuel segments near the bottom of the booster) destroyed the strut attaching the booster to the bottom of the liquid hydrogen tank and burned through the tank itself. The tank erupted into a fireball, and the explosion blew apart *Challenger*. Next, investigators had to understand the reasons for the faulty joint.

In the meantime, President Reagan appointed a special commission to conduct a formal inquiry—the Rogers Commission, named after its chairman, former secretary of state William P. Rogers. The Rogers Commission discovered that NASA had been worried about the booster joints for several months. The specific problem involved O-rings, circular synthetic rubber inserts that sealed the joints against volatile gases as the rocket booster burned. It was believed that the O-rings lost their efficiency as boosters were reused; their efficiency was even more reduced in cold weather. The Rogers Commission further discovered that NASA and managers from Thiokol, suppliers of the solid fuel boosters, had hotly debated the decision to launch, during the night before *Challenger*'s fatal flight.

The Rogers Commission report, released in the spring of 1986, included an unflattering assessment of NASA management, calling it "flawed," and recommended an overhaul to make sure managers from the centers kept other top managers better informed. Other criticisms not only resulted in a careful redesign of the booster joints but also led to improvements in the shuttle's main engines, a crew escape system, modified landing gear, alterations to the landing strip at Kennedy Space Center, and changes in a host of aspects of shuttle operations. NASA originally planned to resume shuttle flights in the spring of 1988, but nagging problems delayed new launches through the summer.

In the wake of *Challenger*'s destruction, other changes occurred. Some realignment would have occurred in any case, since NASA administrator James Beggs, indicted for fraud and later completely exonerated, had vacated the position in December 1985. At the time of *Challenger*'s loss, an interim leadership was in place; in the aftermath of *Challenger*, James C. Fletcher returned to NASA's helm. But the shuttle tragedy colored many subsequent senior management reassignments in NASA, along with a reorganization of contractor personnel. Even though President Reagan authorized construction of a new shuttle for operation by 1991, the existing fleet of three vehicles remained inactive for over two and one-half years, severely disrupting the planned launch of civil and military payloads. For some scientific missions, desirable "launch windows" were simply lost, and other missions, rescheduled sometime in the future, were severely compromised in terms of scientific value. In the case of the Space Shuttle program, NASA had not only stumbled but was left staggering.

New Directions

Although the flight of *Voyager 2* past Uranus and on toward Neptune represented a striking success, it was almost overlooked in the clamor triggered by the loss of *Challenger*. During the next several months, the agency's frustrations multiplied.

In 1986 Halley's Comet made its appearance again, after an absence of seventy-six years. Halley, as the brightest comet that returns to the Sun on a predictable basis, was a valued astronomical performer, and scientists had adequate time to prepare for its reappearance. However, during Halley's dramatic swing across Earth's orbit, many American scientists lamented that no American spacecraft was on a mission to meet it and take scientific measurements. Some satellites launched by the United States were able to make ultraviolet light observations, but only the ESA, Japan, and the Soviet Union had planned to send probes close enough to use cameras—ESA's Giotto probe came within 375 miles of Halley's nucleus. Critics charged that excessive NASA expenditures on the shuttle had robbed America of the resources to take advantage of unusual opportunities such as the passage of Halley's Comet.

In the aftermath of *Challenger,* NASA's hopes for recovery seemed plagued by a rash of additional misfortunes. In May 1986 a Delta rocket carrying a weather satellite was destroyed in flight after a steering failure. One of NASA's Atlas-Centaur rockets, under contract to the U.S. Navy for the launch of a Fleet Satellite Communications Spacecraft, lifted off in March 1987 but broke up less than a minute later after being hit by lightning. During the assessment of the loss, a review board scolded NASA managers for making the launch into bad weather conditions that exceeded acceptable limits. In June three rockets at NASA's Wallops Island facility were being readied for launch when a storm came in. Lightning hit the launch pad and triggered the ignition of all three rockets; frustrated engineers watched the trio shoot off in hopeless trajectories over the Atlantic shoreline before crashing into the sea. In July disaster hit NASA again when an industrial accident on the launch pad at Cape Canaveral destroyed an Atlas-Centaur upper stage, forcing cancellation of a military payload mission.

These embarrassments, along with the brooding shadow of *Challenger,* dulled the otherwise bright successes. Early in 1987 determined launch crews had successfully put two important payloads into orbit. The *GOES-7* environmental satellite went into operation, returning vital information on the formation of hurricanes in the Caribbean. An Indonesian communications satellite, Palapa B 2P, originally scheduled for a shuttle launch, went into orbit aboard a Delta rocket launched from Cape Canaveral. While debate over the nation's space program persisted, NASA continued its space work on several different projects. Taken collectively, they held considerable promise for many areas of both astronautics and aeronautics.

With the resumption of Space Shuttle missions for which special payloads were developed, many in the space science community hoped for a renaissance in astronomical science. Particularly high hopes attended the development of the Hubble Space Telescope. Weighing twelve and a half tons and measuring forty-three feet long, the Hubble Telescope with its 94.5-inch mirror constituted the largest scientific satellite of its era. All ground telescopes are handicapped by Earth's atmosphere, which distorts and limits observations. The Hubble Telescope would permit scientists to collect far more data from a wide spectral range unobtainable through present instruments. The most alluring prospect of the Hubble Telescope's operation lay in its potential to search for clues of other solar systems and gather data about the origins of our own universe. With the telescope in orbit, astronomers expected it to pick up objects 50 times fainter and 7 times farther away than any ground observatory; via electronic transmissions to Earth, the telescope could let humans see a part of the universe 500 times larger than has ever been seen before. NASA press releases speculated about unparalleled opportunities to capture images of primeval galaxies as they started to form at the beginning of time. Fittingly, the Hubble Space Telescope came to be an international enterprise, with the ESA supplying the solar power arrays and certain scientific instruments as well as several scientists for the telescope's science working group.

The Hubble Space Telescope was not the only major effort in astronomy, astrophysics, or planetary research. NASA planned a new family of orbiting observatories, often developed with foreign partners, to probe more deeply into the background of gamma rays, infrared emissions, celestial X-ray sources, ultraviolet radiation, and a catalog of other perplexing subjects. There were also several bold planetary voyages to be launched. In collaboration with the Federal Republic of Germany, the Galileo mission to Jupiter (requiring a six-year flight after launch from the Space Shuttle) called for an atmospheric probe to be parachuted into the Jovian atmosphere while the main spacecraft went into orbit as a long-term planetary observatory. The Magellan mission envisioned a detailed map of the planet Venus; Ulysses (planned with ESA) was designed to explore virtually uncharted solar regions by flying around the poles of the Sun. Closer to home, environmental concerns culminated in a catch-all program dubbed Mission to Planet Earth, an idea that waxed and waned over the next several years. All of these missions were targeted for the late 1980s and early 1990s; creative scientists and engineers were also concocting ambitious projects for the twenty-first century.

In the spring of 1987, NASA made a determined step toward lunar and Martian missions by creating the Office of Exploration to begin planning. NASA's plans for an operational space station, though not crucial for these goals, were nevertheless important, since the station could play a major role in their support.

The first technically reasoned studies of a space station began in the late 1930s, when Arthur Clarke and his friends in the British Interplanetary Society began publishing proposed designs. Rocketry in World War II seemed to make these speculations far less sensational to the postwar generation. In March 1952 the popular American magazine *Colliers* startled some readers but fascinated others with a special edition on space exploration. One of the more dramatic articles featured a space station shaped like a huge wheel, 250 feet in diameter, designed to rotate in order to provide artificial gravity for the station's inhabitants.

During the next three decades, variations of the *Colliers* design and other space station structures appeared in various popular and technical journals. Some early ideas, like the need for artificial gravity, persisted for a long time before finally disappearing (except for special requirements like centrifuge experiments). Others, including modular structures, free-flying "taxis," and a stationary facility for zero-gravity activities, remained staples of space station thinking. With the organization of NASA in 1958, space station planning took on a more practical aspect as part of a national commitment to space exploration. Within two years of its founding, NASA had organized a committee within the Langley Research Center to study the technology required for space stations.

The process of deciding on the design of a space station and its uses consumed more than two decades and several million dollars. A significant milestone was President Ronald Reagan's endorsement of the Space Station *Freedom* program in his State of the Union message in January 1984. Meanwhile, NASA and contractor space station studies proceeded through several variations before NASA designated one design as the "baseline configuration." This structure, which emerged during 1987–88, was scaled down in size because of budgetary constraints and the reduced number of shuttle flights after the loss of *Challenger*. A primary concern was to put a station in operation by the mid-1990s, holding the cost to about $8 billion. At the same time, NASA publicized what it called a phased approach, giving the agency an option for adding several large components once the basic space station was in place.

The American space station initiative included an invitation to foreign partners to share in its planning and operation; refining the details of this partnership engaged negotiators from the United States, Canada, Japan, and the

ESA over the next four years. The toughest negotiations involved ESA. The Europeans wanted to ensure free access to the space station and to guarantee some technology transfer in return for their contributions to station development. The foreign partners also strenuously resisted plans for significant space station activities by the American armed services. The United States and its international partners agreed to limit space station uses to "peaceful purposes," as determined by each partner for its own space station module. The final documents were signed by ESA, Japan, and Canada in September 1988. The United States planned to build a laboratory module and a habitation module for the crew. The Europeans and the Japanese took responsibility for the two additional laboratory/experimental modules; Canada agreed to supply a series of mobile telerobotic arms for servicing the station and handling experimental packages.

Aeronautics

Aeronautical research proceeded along several lines. The Grumman X-29 began flying additional missions to test upgraded instrumentation systems. With air force cooperation, a considerably modified F-111 carried out flight tests using a Mission Adaptive Wing, in which the wing camber (the curve of the airfoil) automatically changed to permit maximum aerodynamic efficiency. With the DoD, NASA launched the development of a hypersonic aircraft, the X-30, tagged with the inevitable acronym NASP, for National Aero-Space Plane. Plans called for a hydrogen-fueled aircraft that would take off and land under its own power. The plane would streak aloft at Mach 25 and would be able to operate in a low Earth orbit much like the shuttle, or cruise within Earth's atmosphere at hypersonic speeds of Mach 12. Its ability to sprint from America to Asia in about three hours encouraged the news media to refer to it as the "Orient Express." A series of developmental contracts awarded during 1986 and 1987 focused on propulsion systems and certain aircraft components; an experimental, interim test plane was several years away.

Other flight research represented a totally different regime of lower speeds and an emphasis on fuel efficiency. Even though jet fuel prices dropped in the mid-1980s, the cost was still five times the 1972 amount and accounted for a significant percentage of operating costs for airlines. For that reason airlines and transport manufacturers alike took intense interest in a new family of propfan engines sparked by NASA's earlier Aircraft Energy Efficiency Program. Using a gas turbine, the new engine featured large external fan blades that were swept and shaped so that their tips could achieve supersonic velocity. This

In 1972, NASA and the FAA embarked on a cooperative effort to develop technology for improved crashworthiness and passenger survivability in general aviation aircraft with little or no increase in weight and acceptable cost. Since then, NASA has "crashed" dozens of GA aircraft by using the lunar excursion module (LEM) facility originally built for the Apollo program. The aircraft are swung by cables from an A-frame structure that is approximately 400 feet long and 230 feet high.

would allow the propfan to drive airliners at jetlike speeds but achieve fuel savings of up to 30 percent. Different trial versions of multibladed propfan systems were in flight test beginning in 1986, with operational use projected by the early 1990s. Although the team responsible for the propfan development received the Collier trophy, a decrease of tensions in global politics plus better utilization of petroleum resources brought about a surge of oil production that eased the energy crisis. The propfan technology went into storage—but remained available should the need arise sometime in the decades ahead.

Investigation of rotary wing aircraft continued, even as the experimental XV-15 Tilt Rotor Research Aircraft evolved into the larger V-22 Osprey, built by Boeing Vertol and Bell Helicopter for the armed services. A joint program linked the United Kingdom, NASA, and the DoD for investigation of advanced short-takeoff and vertical-landing aircraft. Basing their thoughts on the concept used in the British Harrier "jump-jet" fighter, designers began wind tunnel tests of aircraft that could fly at supersonic speed while retaining the Harrier's renowned agility. Eventually, such arrangements led to British participation in the development of advanced combat aircraft like the Joint Strike Fighter scheduled by the U.S. Air Force for deployment early in the twenty-first century.

Several new NASA facilities promised to make significant contributions to these and other futuristic NASA research programs. NASA's Numerical Aerodynamic Simulation Facility, located at Ames and declared operational in 1987, relied on a scheme of building-block supercomputers capable of 1 billion calculations per second. For the first time, designers could routinely simulate the three-dimensional airflow patterns around an aircraft and its propulsion system. The computer facility permitted greater accuracy and reliability in aircraft design, reducing the high costs related to extensive wind tunnel testing. At Langley, a new National Transonic Facility permitted engineers to test models in a pressurized tunnel in which the flow of super-cooled nitrogen replaced the traditional medium of air. As the nitrogen vaporized into gas in the tunnel, it provided a medium more dense and viscous than air, offsetting the scaling inaccuracies that accompanied the testing of smaller models (usually with a wingspan of three to five feet) in the tunnel. Nonetheless, large tunnel models and full-sized aircraft still provided critical information through traditional wind tunnel testing. Additional tunnel facilities at Ames opened new avenues for low-speed testing of helicopters, VSTOL and STOL aircraft and other proposed designs.

Spin-off

As a legacy of its activities during the Apollo era, NASA evolved into an agency with myriad functions. During the peak of Apollo program research in the 1960s, NASA became committed to the spin-off concept—that space technology and techniques have other applications also. There has been a series of organizational efforts to publicize and encourage the practical application of new technologies ever since. The Apollo era's legacy included considerable biomedical information and physiological monitoring systems, developed for manned spaceflight, that were widely implemented in hospitals and medical practice generally. The development of the Saturn launch vehicles prompted widespread improvements in bonding many dissimilar materials, handling and machining exotic alloys, and regularizing cryogenic applications and introduced new approaches in production engineering.

The energy crunch of the 1970s caused NASA to consider ways of transferring to the marketplace its considerable expertise in insulation materials, solar energy, heat transfer, and similar topics. In a different context, NASA developed an entity called the Computer Software Management and Information Center, known by the impressive acronym COSMIC, which is a library of multitudes of software programs derived from multifarious applications developed for use in the space program. Patrons who made the effort to access COSMIC saved valuable time and millions of dollars by using available programs rather than developing new ones or risking serious design flaws by doing without.

These and other activities constituted a significant NASA contribution to economic and commercial development. The "commercialization of space," a theme of President Ronald Reagan's space policy in the late 1980s, promised many more benefits stemming from renewed shuttle missions and an operational space station. Advances in metallurgy, biology, and medicine seemed the likeliest to be realized in the near future. These programs implied more and more reliance on manned flight, a situation that continued to disturb the practitioners of space science, underscoring a dichotomy in the nation's program that had persisted for many years.

In 1980 NASA's budget stood at $5 billion; it was $10.7 billion for the 1989 fiscal year. Manned spaceflight accounted for more than half of that budget, whereas space science accounted for $1.9 billion, or about 18 percent. The space science share of funding averaged about twenty cents of each NASA dollar consistently over the years. Critics of the space program often cited its lean budgets for science and grumbled that so many shuttle flights were scheduled for military missions. This fact, coupled with the need for twenty or more shuttle missions to deliver space station components into orbit, meant that there would be fewer potential space science payloads.

Critics also pointed out that the cost per pound of shuttle missions exceeded early projections by a considerable margin, undercutting the original arguments in favor of the manned launch system. The air force had already, in the early 1980s, begun development of a family of expendable launchers, to reduce costs and provide alternatives to the possibility of a grounded shuttle fleet. Many foreign customers found it economical to rely on the Ariane launch vehicle, operated under the authority of the ESA. NASA itself pondered the use of a new series of expendable launch vehicles to complement the shuttle. Complicating the picture was the potential competition from a new Soviet shuttle vehicle; ESA also had plans for a similar reusable spacecraft. Finally, the U.S. space commercialization policy prompted several U.S. companies to plan a variety of privately designed and built launch vehicles, which would also compete with NASA's own rocket launchers and the Space Shuttle.

NASA followed many new directions in the process of inaugurating the

Space Shuttle era, the space telescope, and a growing roster of international satellites and other spacecraft. Aeronautical research had likewise launched distinctively new shapes into the sky, including remotely piloted aircraft. Following these diverse paths met the demands of many constituencies, although the agency found that diversity placed new pressures on NASA management and on financial resources. Emphasizing spin-off initiatives seemed to satisfy some critics, who still looked for "dividends from space" as a justification for NASA's billion-dollar budgets. In the years ahead, increased reliance on international support in order to fashion ambitious space initiatives became more commonplace. At the same time, international competition in aeronautics required NASA to pay attention to techniques and technologies that would allow the American aerospace industry to hold its own in a global marketplace.

Chapter 9

The Post-*Challenger* Years, 1989–1990s

As NASA prepared to enter the post-*Challenger* era, it did so under a leadership that had changed in demographics and character. In the spring of 1989, Administrator James C. Fletcher (age 69) left the agency for the second time, having served from 1971 to 1977 and again since 1986. Fletcher's principal tasks during his second tenure had been to accomplish recovery for the Space Shuttle program after the loss of *Challenger* and to give NASA a new kind of momentum. In the early 1990s, flight research also demonstrated positive activity. With activities often framed in the context of the cold war tensions of the recent past, several of NASA's programs in manned spaceflight experienced a historic shift. Political developments in the Soviet Union eventually led to its dissolution, symbolically marked by the demolition of the Berlin Wall in 1989. Within three years after that, American and Russian officials signed agreements for joint missions aboard a Russian space station and for cooperation with other European states in building a new international outpost in space.

Change and Continuity

Although the shuttle program resumed operations in September 1989, Fletcher himself admitted some regret that NASA seemed to lack the sort of broad public and political support that had characterized its earlier years. He warned that his successor would inherit a rash of incipient problems. Among other things, NASA expected the departure of many skilled personnel who had committed to stay only until the shuttle resumed flights. Moreover, the veto of a federal pay raise meant that there would be problems in recruiting many qualified people, and new ethics rules to prevent high-level executives from working for certain contractors for up to two years after NASA service also threatened the recruitment of key personnel. Other personnel issues also threatened to create a serious gap in NASA's structure. Between 1989 and 1991, about 70 percent of the agency's senior and middle management were due for retirement. Because of a hiring freeze in the mid-1970s, there was a shortage of managers in the 30-to-45 age group prepared to move quickly into agency leadership. Some observers saw the possibility of restructuring the agency by making field centers into contractor-operated facilities in the future.

These and other issues fell to a young new administrator, Richard Truly (age 51), who was significantly different from NASA's past leaders. Rear Admiral Truly, U.S. Navy, had been serving as NASA's associate administrator for spaceflight when President George Bush tapped him as the new NASA administrator in 1989. Truly became the first person to head the agency who had actually flown in space. He piloted the second shuttle flight in 1981 and commanded a *Challenger* mission in 1983. NASA's top executive traditionally had been a civilian, with a background in civil service or in the private sector. Indeed, the enacting legislation of 1958 specified that leadership of the civilian space agency should not come from the military. Even though Truly retired from the U.S. Navy, Congress had to pass a waiver permitting his appointment.

In the wake of the *Challenger* explosion in 1986, the Reagan administration initiated new policies to take NASA out of the business of large rocket-launched commercial payloads. NASA's final commercial missions lifted off in the autumn of 1989. The potential payoff from NASA's scientific missions was simultaneously underscored by the continuing odyssey of *Voyager 2*, which performed a memorable grand finale as it passed Neptune from August through December 1989. Since its launch in 1977, the amazing *Voyager 2* had cruised outward through more than 4.4 billion miles of space. It arrived on time and at the right place, streaking past at a distance of three thousand miles from Neptune—the first artifact from Earth to visit Neptune in orbit. Although the major encounter began in August, equipment continued to look back at the planet through December, transmitting a series of continuing surprises to delighted scientists. The ocean-blue planet turned out to have jet-stream winds of 1,500 MPH and a collection of six previously unknown moons; cameras also uncovered five rings circling the planet. As an encore, *Voyager 2* relayed extraordinary new information about Triton, Neptune's largest moon. Triton was found to be the coldest body in the solar system (–400°F) and to be dotted with

Attached to the "robot arm," the Hubble Space Telescope emerges from the Shuttle Orbiter cargo bay, lifted up into the sunlight during this second servicing mission, designated HST SM-02.

geyserlike volcanoes spewing nitrogen clouds five miles high. By all accounts *Voyager 2* had been a smashing success.

All this seemed a fitting backdrop for the start of what NASA news releases called a "golden era" of space science between 1989 and 1994. During that period NASA planned thirty-seven major science missions with unusual potential for altering humanity's understanding and view of the universe. NASA expected the Space Shuttle to play a major role in this ambitious new era. With redesign and modification work complete, the trio of remaining shuttles were ready for action again by the autumn of 1989. First off the mark was *Discovery* (STS-26), which lifted off on 29 September 1988 to place the Tracking and Data Relay Satellite into orbit. Two more Space Shuttle launches followed before

NASA began to work off the agency's backlog of major scientific missions delayed by the loss of *Challenger*.

While some spacecraft were beginning their dazzling voyages into outer space, the lives of others were ending. *Solar Max*, launched in 1980 and repaired in orbit during 1984, compiled a distinguished record before succumbing to the inexorable pull of Earth's gravity early in December 1989. As *Solar Max* went down in a fiery reentry, NASA astronauts rehearsed maneuvers to save a different payload in orbit—the Long Duration Exposure Facility (LDEF). The LDEF had been left in orbit by the same crew that had repaired the *Solar Max*. Scheduled to stay aloft for only a year, the LDEF was left stranded in space by shifting priorities and the loss of *Challenger*. The job of snagging and retrieving the eleven-ton satellite about the size of a school bus fell to the crew of *Columbia*, launched 9 January 1990. Because the LDEF was beginning to accelerate into a descending orbit, a sense of urgency surrounded the mission; among its experiments was the study of a variety of materials as candidates for the space station and future spacecraft. After deploying a military communications satellite, the *Columbia* crew successfully shifted orbits and coasted to within 30 feet of the LDEF, where mission specialist Bonnie Dunbar manipulated the shuttle's 50-foot robotic arm to snare it and pull it down inside the shuttle cargo bay.

Other missions seemed to help set the stage for similar operations involving a space station. In the spring of 1993, for example, one Space Shuttle carried a Spacehab module on its maiden flight. Privately funded and developed, the Spacehab rode in the shuttle's cargo bay, snugged up against the crew compartment to provide more than one thousand cubic feet of space. Although the Spacehab quadrupled the available work and storage volume of the shuttle, the rest of the cargo bay remained free for large payloads. On Spacehab's first mission, NASA and industry carried out four biotechnology procedures, plus eleven other experiments in materials science. A lineup of investigators for future Spacehab opportunities verified its potential. Additional shuttle flights continued the interaction with international partners—another parameter of the space station under development. A different mission in 1993 carried the European Space Agency's Spacelab, a much larger module than Spacehab. Funded largely by Germany, it was equipped for extensive experimental procedures, to be carried out by technicians in the Spacelab's shirtsleeve working environment. Two German payload specialists flew along to perform some ninety different experiments. Most of the assignments involved research for ESA, but Japan also consigned a number of projects, and NASA itself assigned a trio of tasks to the flight.

In subsequent Space Shuttle flights, European activities took center stage. One shuttle mission successfully picked up ESA's European Retrievable Carrier (EURECA), originally deployed during a 1992 shuttle flight. During its ten-month stay in Earth orbit, EURECA carried out automated or radio-controlled biological experiments and studies in metallurgy and crystal growth. Later in 1993 a shuttle deployed and recovered a German experimental satellite during the same mission. This was the Orbiting and Retrieval Far and Extreme Ultraviolet Spectrograph, more conveniently known as ORPHEUS. The scientific instrument package for ORPHEUS was mounted on a standardized shuttle pallet to handle either solo German missions or combined missions with NASA payloads aboard. NASA's collaboration with international partners continued to expand.

With these fascinating missions winning their share of international headlines, the new presidential administration of George Bush set a number of long-range priorities for the agency. These were to complete the proposed space station (dubbed *Freedom*), to set up a manned lunar outpost, and to begin plans for the manned exploration of Mars. Interim projects like the Hubble Space Telescope kept NASA and its contractors fully engaged. Along with outward-looking ventures, NASA also proposed a major interagency effort to study the complex interactions of human civilization and the global environment. NASA's ongoing study of the "ozone hole" above the Antarctic typified such projects. All of these concepts presumed extensive European and Japanese collaboration. This growing internationalism was underscored by several new agreements that included closer collaboration with Russia as well.

Missions to Outer and Inner Space

After reaching the planet Jupiter, the Galileo spacecraft also turned in a successful performance. By the middle of the 1990s, a string of successful finales involving lengthy planetary and solar missions helped to brighten NASA's improving image and cheer its international partners. Launched with German participation in 1989, Galileo coasted into Jupiter's neighborhood during 1995. The Galileo mission actually included two different sorts of instrumented components. In 1995, after going into orbit around the solar system's biggest planet, the Galileo probe detached itself and plunged down through the Jovian atmosphere. Fore and aft heat shields protected the probe during its high-speed entry; after some deceleration occurred, a parachute deployed and the probe began data transmissions. Before its inevitable disintegration in the planet's turbulent, gaseous environment, the probe sent back remarkable data on the extent of water there, along with information about dense cloud formations and their chemical composition. The Galileo spacecraft component that remained in orbit continued to send reams of fascinating data—not only about Jupiter, but also about its remarkable moons. Io, for example, displayed new volcanic change since the passage of the Voyager spacecraft seventeen years earlier. Three-dimensional images of Ganymede revealed huge, icy fissures in its surface. Europa yielded hints of remarkable icy oceans beneath its frozen crust. If the hints are accurate, perhaps new life forms might be found there.

As Galileo continued on its orbital analysis of Jupiter's environment, additional reams of tantalizing data continued to pile up. Magellan, launched in 1989, successfully reconnoitered Venus in 1993–94, before burning up in the Venusian atmosphere. Ulysses, planned with ESA and launched in 1990, required a three-year journey before returning data about the Sun's polar regions; its orbital path would bring it back past the Sun again in 2000.

There was one near-calamity in the midst of all this that involved the vaunted Hubble Space Telescope. The Hubble program experienced an erratic developmental history. Its big mirror proved difficult to grind and polish in order to meet stringent qualifications. One of NASA's most publicized projects, it seemed plagued by cost overruns and delays. Hubble's tribulations fed debates in the scientific press about the apparently questionable values of NASA's big, expensive space science projects if they continued to siphon off funds for smaller, less quirky hardware with better chances of successful operations in space. NASA and its contractors persisted, finally clearing the Hubble Space Telescope for launch aboard the Space Shuttle *Discovery* in the spring of 1990. After its successful deployment in orbit, the images transmitted by Hubble turned out to be fuzzy and ill-defined; a crisis management team finally had to report that the mirror was seriously flawed. Hubble, they said, would never produce the clarity necessary for scientific purposes. As one astrophysicist bitterly commented, "This is one of the worst things to happen to astronomy since the Pope strung up Galileo."

Hubble's woes came on the heels of a string of highly publicized disasters going back to the *Challenger* explosion of 1986. News stories repeated earlier reports about launch vehicles that blew up or failed to get payloads into proper orbit, endemic cost overruns, and so on. NASA management struggled to get a handle on these problems and to restore its tarnished reputation.

In the case of Hubble, NASA had planned all along to change or improve some of its equipment in orbit. This now became a do-or-die effort. Late in 1993, a shuttle repair flight effected a rendezvous in space with the ailing Hubble. Astronauts spent several days replacing defective equipment and integrating

An artist's conception of the "International Space Station: Assembly Complete." NASA envisioned the space station as a test bed for the technologies of the future, a laboratory for research on new, high-technology industrial materials, communications, transportation, and medical research. Participants include the United States, Canada, Italy, Belgium, the Netherlands, Denmark, Norway, France, Spain, Germany, the United Kingdom, Japan, and Russia. *NASA Photo HqL-407, 8/94.*

new components designed to raise Hubble's performance to its intended range of specifications. With immense relief Hubble managers reported that subsequent data and images from the space telescope vindicated the controversial program; missions like these also proved valuable in terms of planning EVA work sequences for the planned space station. In 1997, during a subsequent manned shuttle mission to the Hubble Space Telescope, the crew successfully docked the orbiter and then spent five days in space walks to carry out maintenance tasks and to complete upgrades to the telescope's instrumentation. All this patience and additional attention eventually paid off; some of Hubble's pictures were nothing short of spectacular. Photographic images from Hubble continued to mesmerize observers, including one of an awesome collision of galaxies that made the front pages of newspapers around the world and also became the cover for *Newsweek* magazine.

Other initiatives, announced with appropriate fanfare and press conferences, seemingly made their appearance as an effort to snare congressional goodwill and public support. Although eye-catching reports about experimental aeronautics or dramatic launches to far-off galactic regions got most of the headline attention, NASA's new focus on the earth's environment received a growing share of news stories. In actuality, Mission to Planet Earth (MTPE), as the agency introduced it during the early 1990s, was more than just a response to environmental activists. MTPE embraced a diverse range of activities that promised to enlarge our understanding of geophysics, weather, and a host of details about the physical world. The program broadened international collaboration. NASA explained that MTPE gathered data that allowed global citizens to make informed policy decisions, based on new findings about deforestation, oceanography, climate, and "the solid Earth itself."

Among other things, NASA busied itself with studies of the ozone layer. In the early 1990s, mapping spectrometers found record low levels in the midlatitudes of the Northern Hemisphere. Specially equipped satellites evaluated ozone levels over both of Earth's poles and also recorded declining values. Reports like these prompted continuing watchfulness. When equipment aboard one of NASA's ozone recording satellites lived out its fourteen-year lifetime, two more satellite data recorders were put into orbit, one of them aboard a Russian Meteor-3 satellite. In other ventures, Landsat satellites kept track of deforestation and its many effects in the Amazon River basin. Volcanoes came under intensive study, revealing significant effects following heavy eruptions. One extensive observation program, carried out by aircraft, used infrared instruments to evaluate volcanic action along portions of the highly active Pacific Rim. The studies focused on sites in Russia's Kamchatka Peninsula, north of Japan, and featured unprecedented cooperation with Russian scientists. Additional projects within the region continued long-standing oceanographic observations. Specialists from CNES, the French space agency, became close collaborators in some studies, which predicted El Nino effects for 1993–94. More than a dozen other major atmospheric investigations were part of satellite launches as well as shuttle missions covering everything from midwestern summer floods and California brush fires to measurements of the Greenland ice skirt.

New Directions in Management

Elected in 1992, the new administration of William Clinton endorsed the incumbent NASA chief, Daniel Goldin, who had been appointed by the former president, George Bush, in April 1992. Goldin had a background of extensive corporate experience with aerospace contractors like TRW (Thompson-Ramo-

Concern about fuel shortages during the 1970s and 1980s led to many studies for more efficient propulsion systems. The Advanced Turboprop Program at NASA's Lewis Research Center received the 1987 Collier Trophy award.

Wooldridge). He wanted to revamp NASA's management, get better control of its budget, and revive the agency's spirit. He found that one symbolic change produced immediate results. During 1975 NASA had implemented a program to update its image, replacing its traditional round logo and flight symbol with modern curved typography that simply spelled NASA. Critics referred to the curvilinear logo as "the snake"; most personnel detested the change. When Goldin made a series of goodwill tours in 1993, not long after his appointment, numerous NASA employees told him that a return to the old logo would help boost sagging morale. This switch became one of his earliest—and most popular—decisions. Also, NASA added a new orbiter, Endeavor.

As NASA continued to experience pressure to reduce costs, its top management struggled to fund current programs and still provide adequate budgetary support to R&D programs needed for future projects. In 1994 Goldin recruited a panel of experts to formulate a new strategy. After several months of analysis, the advisory group recommended that NASA assign virtually all day-to-day operational functions to a centralized aerospace contractor entity. NASA itself would continue R&D activities at its centers and maintain its primacy in astronaut training, launch operations, and mission control. This operational streamlining would result in lower contract costs and yield the R&D moneys that NASA desired. In 1996, after a year's analysis and negotiations, the initial contract went to a joint venture, the United Space Alliance, formed by Lockheed-Martin and Boeing to maintain the Space Shuttle. Subsequent awards went to major contractors to consolidate and manage ground operations for other programs, and other NASA centers adopted similar contractual arrangements. As Goldin emphasized, the shift symbolized a major turning point for the agency.

The Russians Are Coming

The early 1990s were full of dramatic political changes that reshaped the aerospace community. Following the breakup of the Soviet Union, the United States knew it had to cope with inevitable contraction in the military sector. Mergers of major aerospace corporations in the United States reshaped the contours of the entire industry. By the mid-1990s, companies like Convair, McDonnell Douglas, and others had been subsumed by other corporate entities: for example, Convair became part of Lockheed-Martin, and McDonnell Douglas went to Boeing. International ventures became more commonplace, with an added element of historic drama. Russia—long an opposing superpower—now became a key partner in NASA's plans to orbit a large space station. A drawn-out sequence of negotiations paved the way for this striking turn of events, which included a series of Space Shuttle missions to dock with Russia's space station *Mir*.

In the aftermath of American success in the manned lunar landing race, the Soviet Union had taken an expedient route in a different direction. During the competition for primacy in manned lunar landings, the Soviets' big boosters failed to complete a successful launch. On the other hand, launch vehicles carrying cosmonauts into Earth orbit continued to rack up an enviable record, giving the Soviet manned space program an increasing level of expertise. So the USSR reoriented these skills to step up work toward long-duration space stations. *Salyut 1,* the first example, arrived in position during 1971. That same year, a three-man Soyuz spacecraft successfully docked. The crew spent a busy three weeks in orbit; then a tragic accident during reentry took their lives. After nearly three years of reassessment and design, a new Salyut lifted off, soon followed by a Soyuz carrying a pair of cosmonauts who docked for a 15-day mission. The following years saw a flock of Soyuz flights to other new Salyut stations, with some crews staying up to seven months and welcoming other space travelers during their stay. One veteran was Valery Ryumin, who—with Lieutenant Colonel Vladimir Lyakhov—spent a challenging 175 days in orbit during 1979. Seven months later, in 1980, an injury to another cosmonaut sent Ryumin back into space for another 184 days. This flexibility, plus the dozens of experiments performed and the numerous in-flight repairs and ad hoc docking maneuvers successfully carried out, clearly demonstrated the maturity of the Soviet space program.

The USSR already had more ambitious plans under way. The new, larger station called *Mir* (translated "peace") was launched 19 February 1986. Its modular design facilitated later additions for experimental work, solar arrays, and additional systems, with room for two to three crew members. Its long career eventually included the first American spaceflights to a Russian space station.

Agreement with the Russians to participate in the International Space Station (ISS) program evolved from a mixture of self-interest and international goodwill in the adventure of space exploration. The latter set the context; the former provided the impetus. The Russians had compelling expertise in manned space operations, especially in long-duration missions, as demonstrated by Salyut and *Mir*. Financial realities in Russia required that any activities beyond *Mir* be undertaken in collaboration with the United States and western Europe. European cooperation in the ISS venture also demonstrated the American need for international budgetary support. Using Russian technology and skills made sense at a time when congressional budget hawks

In the energy-conscious era of the 1970s, NASA's operational experience found many new applications. Prototype solar panels provided power to these buildings.

seemed intent on shaving more and more from NASA's appropriations, particularly in the case of the ISS. Its opponents argued that big manned projects, requiring many shuttle flights, depleted NASA's budgets for many space science missions and for aviation research.

On 5 October 1992, following many weeks of discussions, NASA and the recently organized Russian Space Agency (RSA) hammered out an Implementing Agreement on Human Spaceflight Cooperation. At the heart of it was the agreement to begin training Russian cosmonauts for Space Shuttle missions beginning early in 1994. Later in 1992 NASA administrator Goldin and Yuri Koptev, his RSA counterpart, met in Moscow to finalize arrangements for American astronauts to train for duty aboard *Mir*. The Russians also negotiated to receive American funding for certain hardware and for training purposes. For an uncertain Russian economy, the U.S. dollars amounted to a significant benefit.

All of this codified plans for Phase I. Phase II called for Russia to design and fabricate major components for the new international space station. Russian and American components joined in space offered the prospects for an orbital research facility, tended by on-board crews from Earth, much sooner than was otherwise likely. This was to be followed by Phase III, involving the construction of a full-scale space station with international participation along the lines already hammered out by NASA, ESA, and other partners. Preliminary plans projected a ten-year lifetime for this major research facility in Earth orbit. With the agreement of its existing international partners, NASA carried the ball during subsequent negotiations that painstakingly worked out details of Russian participation and grappled with technical interfaces, along with a host of ancillary operational issues.

While all these arrangements unfolded, NASA and RSA dealt with immediate issues involving the *Mir* Space Shuttle flights. Between 1994 and 1997, contractual arrangements called for Russia to receive roughly $100 million per year—the American subsidy for the requisite training and hardware required during Phase I. The funds also covered certain elements associated with Phase II, the interim orbital facility. At the same time, some significant subsidiary documents finalized NASA relationships with the Russian State Committee for the Defense Branches of Industry, known by its jaw-breaking acronymic name of GOSKOMUBORONPROM. Almost forgotten in the rush of the more publicized arrangements with the RSA, these negotiations with the Russian defense industry—a bulwark of Soviet truculence in cold war confrontations—surely demonstrated the amazing relationships of the post-Soviet era. Through a Joint Working Group, Russians and their American opposites began a series of far-reaching accommodations that stretched beyond the immediate boundaries of cooperation in space to include aeronautics as well. Catalogued in a 1993 NASA news release, the Joint Working Group agreed to implement collaborative work in eight program areas: aerodynamic transition and turbulence, composite structures and materials, chemically reacting flows, thermal protection system materials, environmental concerns in aviation, hypersonic technologies, experimental test facilities, and advanced aerospace materials. As the NASA news release noted, "This agreement initiates a new era of cooperation with Russia."

By the time these historic events concluded, high-level decisions by NASA, the president, and Congress also restructured the original Space Station *Freedom* architecture. The final design for the ISS featured a 290-foot-long truss assembly with four pairs of solar arrays at each end. The solar array "wingspan" measured 361 feet across. Midway along the truss structure was a grouping of separate modules that provided living space, laboratories, and docking ports, along with additional solar arrays and thermal control system radiators. Normally, a crew of seven astronauts would live and work within this complex, which had an atmospheric pressure of 14.7 pounds per square inch, similar to that on earth. Designers put the average altitude of the ISS at 220 miles at an inclination of 51.6 degrees to the Equator; the projected weight came to about 924,000 pounds. Although downsized in comparison to earlier proposals, the ISS remained a target of disgruntled groups. At NASA Headquarters, managers felt challenged by a continuing barrage of criticism from Capitol Hill, independent agencies, and the scientific community. According to vocal skeptics, the space station had already cost too much, lagged behind predicted milestones, and was diverting NASA's attention from productive, cost-effective development of other hardware for space research. Defenders countered by citing a catalog of new plans for space exploration through the 1990s and into the twenty-first century.

New Dimensions in Flight Research

During the continuing flurry of debate about the direction of NASA budgets and the perennial concern about the possible decline of aeronautical research, NASA's various centers continued to advance the state of the art. During 1994, in a move that buoyed the morale of the aviation community, Dryden reclaimed its status as Dryden Flight Research Center—a full-fledged operational site. The nature of its experimental flight testing continued to evolve in ways that would have perplexed old NACA personnel. Partly because of tight finances, but also because useful hardware remained available, NASA often made striking use of equipment left over from prior experimental programs. And the theme of international collaboration—so pronounced in space station initiatives—also surfaced in the arena of flight testing.

During the 1980s the air force had radically modified a pair of F-16 fighters, equipping them with entirely different wings that featured a delta-shaped planform with a distinctively shaped leading edge. Designers described the resulting F-16XL as having a "cranked arrow wing." After finishing its high-speed analysis with these planes, the air force prepared to scrap them. NASA's Dryden facility alertly negotiated a loan of this pair in 1988 and further modified them for investigations into supersonic laminar flow control. At the time, Rockwell had an interest in this field and agreed to build a glovelike test section to fit over a wing of one of the planes. Made of titanium, the glove had a perforated surface and a system to suck in air across this surface as a means to enhance smooth laminar flow during high-speed flight. A follow-on research series included support from additional manufacturers such as Boeing and

The "Return to Flight" launch of the Space Shuttle *Discovery* and its five-man crew from Pad 39B at 11:37 A.M. 29 September 1988, as *Discovery* embarked on a four-day mission.

McDonnell Douglas. The other F-16XL carried out further analysis of airflow patterns in certain flight regimes.

All this information contributed to NASA studies in what was called the High Speed Civil Transport (HSCT) Program. The F-16XL research flights continued into the mid-1990s. Results from the HSCT investigations might not affect airline transports for some years, but the flight sequences harvested a rich source of data. Further, the association with manufacturers of transport aircraft proved to be educational for Dryden's personnel, who had worked mostly with counterparts from the military aircraft sector. Both NASA and industry team members gained insights into operational requirements and flight procedures that neither had understood until their partnership in the F-16XL flight tests.

Other aerial investigations related directly to military interests. For example, the High Alpha Research Vehicle (HARV) used a McDonnell Douglas F/A-18 to probe the mysteries of the "stall barrier" that compromised the performance of combat aircraft performing at low speeds or assuming a high angle of attack.

Beginning in 1987, the initial series of HARV missions used yarn tufts, dye, and smoke to follow airflow patterns over the F/A-18 in steep flight angles. A second series of flight tests included special-alloy vectoring paddles installed on the plane's engine exhaust nozzles. Moved in unison, they vectored the engine's thrust direction so as to augment pitch and yaw movements during flight maneuvers. Much of this work helped in the correlation of wind tunnel experiments and computational simulations. A third phase investigated high angles of attack with the plane having retractable nose strakes. The data promised useful insights for the design of high performance aircraft in the future. Other HARV experimental flights continued through the 1990s. And, although this was not its original intent, the HARV program generated interest in maneuverability and control by means of thrust vectoring.

A follow-up program ensued very quickly—the development of the X-31 research aircraft. Its international features made this project a special milestone in the evolution of NASA's X-series of experimental flight hardware. The venture began with German aerodynamicist Wolfgang Herbst. In the event of a European air war, Herbst noted, military engagements would logically take place within a constricted sector owing to the limited area of European air space. In this environment maneuverability played a significant role; thrust vectoring at high angles of attack would be especially desirable. But Germany had neither the funds nor the flight test facilities to carry out an experimental program.

Deutsche Aerospace, Germany's largest aero company, along with the German Federal Ministry of Defense, approached the United States. In America, the Defense Advanced Research Projects Agency was definitely interested, along with Rockwell International, one of America's leading aerospace contractors of the era. Eventually, all of these parties were joined by the U.S. Navy, the U.S. Air Force, and NASA. Out of this heterogeneous mix came the X-31. The plane's delta-shaped wing was designed and built in Germany, along with the thrust-vectoring system and much of the flight control laws. Rockwell and other U.S. sources supplied the fuselage and other components. The first of two X-31 planes flew in 1990.

The X-31 venture did not proceed with unanimous NASA approval. Because its agenda dealt specifically with military advantage for a nation other than the United States, some NASA personnel were unhappy to see it at Dryden. Eventually, most of them seemed to agree that the aerodynamic knowledge gained from the program made it worthwhile. The process of melding the several participants into a coordinated research group did not always proceed smoothly. Working out a successful team provided additional lessons in management procedures. Significantly, the thrust-vectoring technology became a promising new avenue of research, and NASA continued to share its subsequent investigation.

In addition to research benefiting military and commercial aircraft, NASA followed additional work in the civil field that promoted the future of the light plane (general aviation) industry. Even though the dollar value of light plane sales (about $8.5 billion in 2000) did not come close to the tens of billions expended for airliners, their sheer numbers (220,000) compared to the scheduled airline fleet (5,000) made them a strong presence in U.S. air space. Moreover, light plane export sales helped create a climate of market acceptance for other American-made aircraft. Accordingly, NASA evolved several R&D programs for this lively sector of aviation, called AGATE, for Advanced General Aviation Transports Experiment. Begun in 1994, AGATE started off with a $100 million budget, an additional $40 million or so from partners in industry, and a lifetime of eight years. Basically, the goal was to find ways of building smaller, cheaper, safer, and easier-to-fly personal airplanes.

NASA aimed the AGATE program toward potential customers in suburban and rural communities with no airline service and customers from large urban centers whose businesses took them to such destinations. Typical flight plans might cover anywhere from 150 to 1,000 miles. Heartened by federal legislation in 1994 that protected manufacturers from punitive lawsuits, allowing builders to reduce costs and attract investment, the general aviation industry still needed to cut the price tags of its planes, enhance safety, and take advantage of advanced technologies. NASA's Langley Center took the lead in several areas of engineering that led to a spectrum of valuable improvements. One new plane, from the Cirrus Design Corporation, in Minnesota, incorporated several of these design gimmicks. The four-place Cirrus SR20, delivered to its first customers in the spring of 1999, featured a single control stick, which replaced separate elements like foot pedals and a control wheel to deal with roll, pitch, and yaw. Using NASA's research studies, designers of the Cirrus incorporated several notable conveniences, such as the flight display panel in the cockpit that could call up images of specified airport runway layouts. For the fuselage and wings, a composite structure reduced labor costs and provided additional safety in terms of durability and resistance to lightning strikes in flight. The Cirrus also incorporated specially designed wings that reduced drag to a minimum and also resisted out-of-control spins. But its most unique hallmark featured an integral parachute that deployed in case of some major failure that might lead to a crash. Instead, the Cirrus and its passengers floated down to earth beneath the parachute canopy. Although the price tag for a Cirrus totaled more than $171,000, NASA hoped that advanced production techniques and rising sales in the future would bring down unit costs for the Cirrus as well as other aircraft.

NASA turned toward the twenty-first century with a renewed vigor. True, budget limitations continued to be a major concern. But the ISS symbolized a marquee program, one that NASA hoped would stimulate interest in space. As some writers pointed out, the nation's early enchantment with the idea of rockets, space ships, and galactic exploration seemed consistent with America's frontier heritage. During the postwar decades, a rash of science fiction movies, especially epics like *Star Wars* and television sagas like *Star Trek* abetted romantic notions about space travel that eventually began to lose some allure. The proposed space station would only accommodate a few a people at a time, and only if they possessed professional credentials far beyond the reasonable expectations of most citizens. On the other hand, thousands of personnel followed satisfying careers with NASA and its contractors, believing that the continuing exploration of the cosmos involved an honorable endeavor in the name of humanity. For those in aviation, aeronautical research involved practical, life-saving issues of airline operations as well as dramatic work in unique areas such as vectored-thrust aircraft. Investigations like these kept additional thousands of aerospace personnel engaged in the pursuit of improved aircraft and safer flight.

Chapter 10

Toward Century 21

During the decade of the 1990s, NASA launched several different broad-scale initiatives, composed with some sort of overarching theme. Budgetary constraints became even more of an issue. Consequently, projects within these broad schemes were often designed to satisfy specific requirements for realizing larger goals, often in concert with international partners in space. In this context, increasingly sophisticated multinational ventures involving planetary probes and collaboration in space science became more commonplace. American astronauts joined their counterparts from Russia aboard space stations in orbit. Aeronautical projects fielded solar-powered aircraft, evaluated systems for next-generation airliners, and tested aerospace hybrids designed to replace an aging fleet of space shuttles sometime in the future.

A Map for the Road Ahead

NASA convened a government-industry team of scientists, engineers, and managers to outline what the agency called the Roadmap for the coming century in the area of space exploration. Organized in 1996, the Roadmap group and its several panels assembled an aggressive list of space exploration projects up to the year 2015. The Roadmap deliberations considered the realities of rapidly fading Russian space budgets, along with a general tightening of moneys from European and American sources, especially in light of escalating costs for the launch and operation of the planned International Space Station. As a result, the large, robust unmanned spacecraft—Galileo, Hubble, and others—were to give way to hardware of more compact architecture, with more advanced computer systems, generally fewer sensors, and leaner budgets. None of this precluded adventurous projects. Even before a Martian meteorite titillated the scientific community with its hints of simple life forms, the search for galactic life and the origins of life itself had been a part of Roadmap deliberations. Some of the other goals were to gain a better understanding of the origins and evolution of our solar system, to find opportunities for human colonization elsewhere in space, and to study aspects of the conditions for climatic change on earth and for impacts by asteroids.

A useful exercise, Roadmap reflected NASA's continuing commitment to adventurous programs. In a cost-conscious federal bureaucracy, the pursuit of adventure often proved difficult. Accordingly, the agency prudently worded its blueprints for the future. Making the transition from the twentieth century to the twenty-first, NASA's policy mandarins crafted a broad agenda that satisfied a tradition of pushing the envelope of flight technology and space exploration and at the same time responded to practical political issues, environmental concerns, and close-to-home economic values in the form of spin-off benefits.

During NASA's early decades, cold war rivalries spurred much of the agency's work in aeronautics and in space projects that demonstrated a sort of "one-upmanship" in rivalry with the Soviet Union. These features had long characterized an annual compendium titled *Spinoff,* a glossy, magazine-style publication filled with color illustrations. By the late 1990s, this approach had plainly changed. In the 1997 version, *Spinoff*'s editors bluntly declared a different emphasis. "Relevancy to NASA's ultimate stakeholder—the public—is top priority in a budgetary climate that dictates tough choices among many opportunities." In 110 pages of discussion, less than one-third of the text covered the achievements of Headquarters and field centers; about one-tenth explained the agency's cooperative organizations for technology transfer and commercialization; over half focused on specific spin-off technologies and their commercial benefits. In *Spinoff 97,* an overlay of four sweeping themes guided NASA's activities. A number of prior efforts were subsumed under this new master plan, and newer initiatives unmistakably carried its imprint. Significantly, given the growing emphasis on commercialization, the four primary thrusts took the name of Strategic Enterprises. One or more centers, as well as JPL, fit into each enterprise: Mission to Planet Earth (Goddard); Aeronautics and Space Transportation Technology (Ames, Langley, Lewis, and Dryden); Human Exploration and Development of Space (Johnson, Kennedy, Marshall, and Stennis); and Space Science (JPL). NASA Headquarters operated as "corporate headquarters," providing organizational structure and management leadership for the individual centers. Clearly, several earlier initiatives were meshed into the Strategic Enterprise scheme, such as the AGATE effort in general aviation, the MTPE activities, and the various dimensions of the Roadmap plan. Fortunately, despite bureaucratic barriers, the spirit of adventure prevailed. In the case of Roadmap, several efforts fairly crackled with technological bravura.

As Roadmap projects began to take shape, refinements to the catalog estimated anywhere from 30 to 40 missions and also noted certain areas of requisite R&D in order to develop needed technologies. For example, one whimsically named mission, Stardust, required a new substance eventually dubbed aerogel. Stardust, eventually launched in February 1999, had the job of intercepting Comet Wild-Z early in 2004, flying within ninety miles of the two-and-a-half-mile-diameter ice ball. The comet carried the name of the retired astronomer who discovered it, Paul Wild, who journeyed from Switzerland to Cape Canaveral to see Stardust lift off. The NASA mission marked the first attempt to collect material beyond the moon and bring it back to Earth. Whizzing by the comet, Stardust was designed to capture tiny fragments about the size of salt grains, using a pop-up collector covered with special glass foam—the aerogel stuff. A return capsule then detached and zoomed back to Earth for a landing in the Utah desert in January 2006. Paul Wild, who would be eighty years old by then, planned to be there to celebrate. Another proposal outlined a trip to Comet Tempel 1, about 233 million miles away, to shoot an instrumented harpoon into the comet as an anchor and drill a sample of the comet's frozen core; data would then be relayed back to Earth. Additional excursions into space included one to Europa, Jupiter's moon, to deploy an automated submersible for cruising around Europa's frigid and slushy oceans. Others included probes to visit Pluto (a thirteen-year one-way journey) or visits to the Sun. Any Martians could expect a number of future visitors, with at least nine separate spacecraft scheduled to drop in early in the twenty-first century.

In a nostalgic coda for spacecraft, *Voyager 1* surpassed *Pioneer 10* as the manmade object to achieve the deepest penetration into space. By 2001 it had been coasting outward through the galaxy for about fourteen years. In case it met with any intelligent forces somewhere in the cosmos, *Voyager 1* carried the recorded sounds of human greetings in fifty-five languages.

From *Mir* to ISS

As for manned spaceflight close to home, the Russians and the Americans continued to expand *Mir*'s capabilities in Earth orbit. The first shuttle arrived in

July 1995 (although the first U.S. astronaut to join *Mir*, Norm Thagard, had arrived aboard a Russian Soyuz spacecraft in March). During an *Atlantis* flight to *Mir* in November 1995 (Mission 74), the shuttle took along some additional solar arrays and a new docking module. All told, the shuttle off-loaded more than 2,000 pounds of food, scientific paraphernalia, and other supplies such as oxygen, nitrogen, and water. On its way back home, the *Atlantis* carried some 800 pounds of science hardware and miscellaneous samples from experiments concluded aboard the space station. Several shuttle flights to *Mir* called for a docked phase lasting five days, and some U.S. astronauts spent as much as five months aboard *Mir*. Shannon Lucid began this phase with her arrival during Mission 76, which also included the first extravehicular excursions by American personnel from *Mir*—in this case, Linda Godwin and Rich Clifford. In this way American personnel gained the long-term experience and the extravehicular experience required for future ISS missions.

In the meantime Phase I operations progressed as hoped, with American crews sharing life with the Russians aboard their space station. Nine shuttle missions to *Mir* involved seven cosmonauts (in addition to other international crew members) and seven NASA astronauts; they spent a total of 975 days in orbit. Not all reports of the *Mir* experience were upbeat: an on-board fire and assorted cooling problems occurred, and an air purification system failed. During one Russian resupply mission, the arriving spacecraft plowed across a solar panel and rammed into the science module, causing an air leak. The U.S.-Russian crew had to seal off the damaged module and write off a group of science experiments. Following a few space walks to make repairs and the arrival of improved computer hardware, life on the Russian space station settled down. Some American critics dismissed *Mir* as a "bucket of bolts" and referred to its various incidents as "Mir-haps." Nevertheless, during the Phase I period, a number of management challenges were met and resolved, and a significant start was made in merging Russian and American space technologies.

The series of shuttle flights to *Mir* proved beneficial in several ways. For one thing, they created the opportunities for more than three dozen experiments carried out by the combined U.S.-Russian crews. A good number involved the life sciences, but a variety of tests yielded significant data that helped engineers design and build a safer, more efficient international space station. One series provided information for enhanced structural integrity. Others examined electrical power stability as well as potential contamination of the ISS from the plumes of thrusters that were needed to keep the station in its proper attitude. These joint missions offered invaluable opportunities for cosmonauts and astronauts to learn about each other's regimens during manned spaceflights and

A view of the Space Shuttle *Atlantis* departing *Mir* Russian space station. This image was taken during the STS-71 mission by cosmonauts aboard their Soyuz TM transport vehicle. A view of Earth appears as backdrop.

to work up joint procedures for the demanding task of assembling the ISS in orbit.

Phase II covered the years 1998–2000 and emphasized deployment of initial elements of the ISS as well as early occupation by international crews. First up, in November 1998, was the *Zarya* ("dawn") module, financed by the United States, built in Russia, and lifted into orbit by a Russian booster from Baikonur, the Russian launch complex situated in Kazakhstan, in central Asia. Early in December a Space Shuttle crew from KSC carried the *Unity* module to orbital rendezvous. The shuttle's international crew checked out various connections and ISS systems, including twenty-one hours of EVA to mate the first two elements. Meanwhile, Russia continued work on the ISS Service Module, scheduled for launch during 1999. By the year 2000, the Phase II schedule called for a basically U.S.-Russian station suitable for a permanent crew of three space travelers. The timetable for Phase III, covering 2000–2004, brought together additional components and hardware from European, Japanese, and other international sources to complete the ISS for planned operations with a crew of seven resident astronauts.

The ISS presented a variety of problems. For one thing, the sixteen nations involved used twelve different languages. An agreement to use English as the standard language eased that issue, and collaboration between America and other countries proceeded relatively well, based on many years of prior commercial and military relationships that had often involved aerospace technology. For the Russians it was different. Despite occasional cooperative experiences, like the Apollo-Soyuz linkup in 1975, the United States, as principal partner and manager, had developed a cold war–era aerospace industry having little or no interaction with the adversarial Soviets. The Soviets had done the same. Both countries had sprawling, entrenched aerospace bureaucracies. The Russian approach to spaceflight and its "technical culture" differed markedly from the American tradition. Fewer people in key positions knew the others' language—and the Russian Cyrillic alphabet was totally different. And then there was the money problem. There was a flow of U.S. dollars into the Russian ISS effort, but the Russian Space Agency (RSA) found itself starved for funds from its own government. By early 1999 the RSA had received only $20 million of the $320 million promised by the Russian government. This had a serious impact for the Russian contractor, Energia, which had responsibility for the Service Module, a crucial component needed to man the ISS.

The issue did not seem so important at first, when the propulsion-storage module, *Zarya*, went through production pretty much on schedule, funded entirely by the United States. The Service Module, funded by Russia, which was already lagging, then dragged on even more slowly, despite efforts by Energia to cope with issues that began to cascade. A massive piece of equipment, the Service Module provided a key element for early occupation of the space station. The Service Module not only contained the living quarters for future crews of astronauts and cosmonauts but also housed the equipment and paraphernalia required for the crew's life support systems while in residence. The first module went into orbit in November 1998, the second module docked with it the following month, and a third mission to the station during June 1999 delivered essential supplies and equipment to prepare the station for operational life. But in the process, follow-on flights experienced delays caused by the Service Module's late delivery. Its launch date was optimistically scheduled for

Seen against white clouds and ocean waters, the International Space Station (ISS) moves away from the Space Shuttle *Discovery*. The U.S.-built *Unity* node *(top)* and the Russian-built *Zarya* module (with the solar array panels deployed) were joined during a December 1998 mission.

late 1999, some eighteen months behind schedule. Consequently, the first resident crew did not expect to move in until sometime early in 2000.

If the Service Module fell further behind, the *Zarya* alone could not keep the station components in orbit; orbital decay required deliveries of more propellants; space allocated for other components consequently disappeared; ISS completion time stretched on, requiring interim funding—and so on. Just to make sure, NASA won supplemental appropriations from a nervous U.S. Congress to find some sort of backup for the problematic Russian Service Module. Luckily, NASA found a workable compromise at the U.S. Naval Research Laboratory, which had been supplying a large payload module used on Titan launches. During 1999, when other problems plagued the launch schedule for the Russian Service Module, NASA's stand-in hardware from the U.S. Navy looked like a shrewd, strategic decision.

Fortunately, work on the Service Module improved, and the schedule picked up; christened as *Zvezda* ("star"), the errant module docked with the station in July 2000. For Russia it was also a money-making venture. *Zvezda,* carried aloft by a three-stage Krunichev Proton booster from Baikonur, sported the distinctive corporate insignia of Pizza Hut, the American restaurant chain, which paid the Russians about $1.2 million to get its logo launched into space. Additional shuttle missions delivered supplies and outside attachments to allow follow-on crews to activate the station and to install external equipment required to move hardware, such as solar arrays, as the station expanded. Another flight delivered the *Destiny*, a laboratory module supplied by the United States. In March 2000 the first international crew of three, dubbed Expedition One, arrived via a Russian Soyuz from Baikonur to establish a long-term human presence aboard the ISS: William Shepherd from NASA, along with Yuri Gidzenko and Sergei Krikalev, Russian cosmonauts.

Observers in the United States maintained that the long delays in ISS schedules could be traced directly to Russia's endemic financial problems. These budgetary issues often created touchy diplomatic relations, since NASA and congressional figures alike chafed at stories about Russian expenditures to keep the aging *Mir* space station in orbit. Several times, NASA officially announced its concern that Russian efforts to keep the fifteen-year-old *Mir* aloft siphoned essential finances from the Service Module program. Somewhat pointedly, at a news conference announcing the Service Module's belated shipment to Baikonur, the chief of the Russian space agency told an audience that *Mir* was not going to be de-boosted from orbit. If nothing else, he declared, funding to keep *Mir* in orbit could be borrowed. Moreover, Russian engineers were developing a plan to keep the *Mir* going without a crew aboard. There were reports of carrying private tourists into orbit and providing unique but expensive voyages through space. But last-ditch efforts to salvage the former Soviet station failed; a final crew departed in June 2000, and *Mir* ultimately made a controlled descent to destruction over the Pacific Ocean in March 2001.

Eventually, a second crew had occupied the ISS, and Pizza Hut folks scored another corporate first by hiring the Russians to deliver "the world's first pizza to be delivered and eaten in space." The Russians, who had negotiated an agreement with a wealthy American investment banker for a trip aboard the aging *Mir,* now decided to use an extra seat aboard a Soyuz supply mission to the ISS by selling it to an American tourist in space. Dennis Tito, a former mission trajectory specialist for NASA, held a contract to pay the Russians $20 million for his once-in-a-lifetime getaway vacation. As the truculent Russians insisted that Tito make the trip, an international brouhaha ensued. NASA adamantly opposed the venture, and members of the sixteen-nation ISS consortium questioned the propriety of one country's profiteering at the expense of others by selling tickets to tycoons to visit an international scientific establishment in Earth orbit. But the Russians forged ahead; Tito completed his eight-day junket in May 2001.

Mars: Past, Present, and Future

Mars, the mysterious "red planet," came under intensive scrutiny. A series of NASA, foreign, and joint projects scheduled from the last years of the twentieth century into the early years of the twenty-first were geared to improve our understanding of the planet and to prepare the way for future manned missions. Running through all of these efforts was a common thread—the search for water. Most scientists concluded that water was once a feature of the Martian surface. What happened to it? What were the consequent implications for life on Mars, past and present?

In 1993 the Mars Observer, one of the first of a new series of spacecraft to set out to the red planet, attracted scientific attention as the first U.S. mission to Mars in seventeen years. It turned out to be a major disappointment, falling silent only a few days before the programmed entry into its Martian orbit. Because the spacecraft had been scheduled to provide a detailed map of the Martian surface and gather atmospheric and weather data for a full Martian year, the lost information created even greater hopes and expectations for subsequent Mars missions.

Interest in Mars perked up some years later during analysis of a meteorite credited with a Martian origin. Investigators found some intriguing material

packed in minute crevices and conjectured that it was microbial. Expectation for life on the red planet ticked sharply upward. While the media was still savoring this news, NASA's diminutive Martian lander, called Sojourner, began a series of widely publicized excursions around the Martian surface.

During 1997 two new Mars missions departed from Kennedy Space Center: Mars Pathfinder and Global Surveyor. After establishing a precise orbit around Mars, the Global Surveyor's task involved data collection about the Martian landscape, especially its polar caps and the intriguing lines that looked like river channels. But Pathfinder was scheduled to arrive first, landing on 4 July 1997. The most fascinating element of Pathfinder was its free-ranging rover, appropriately named Sojourner, equipped with an onboard camera. As Pathfinder settled on Mars, 141 million miles from Florida, millions of viewers on Earth watched the proceedings through its camera. Shortly after, Sojourner jauntily rolled off to explore assorted rocks and other features of the Martian surface. The rover stood about 3 feet high, had a length of 4 feet, and weighed twenty-five pounds. With its six paired wheels, exposed wiring harnesses, and angular camera, Sojourner looked somewhat like a Rube Goldberg creation. Pathfinder's camera followed the rover as it nosed about its immediate area; both craft posted images to the Internet, where more than 500 million hits accumulated during July. But Sojourner eventually fetched up against a boulder and could not manage to disengage itself. Nonetheless, the data from the Martian probes proved to be a rich lode of information—and a major public relations success for NASA.

International scenarios characterized many Mars probes, as evidenced in a series of planned Martian missions early in the twenty-first century. The European Space Agency's Mars Express, Europe's first venture to the red planet, was scheduled for liftoff in 2003 aboard a Russian Soyuz booster. Its payload

This is the first contiguous, uniform 360-degree panorama (originally in color) taken by the Imager for Mars Pathfinder *(IMP)* over the course of sols 8, 9, and 10 (Martian days). Different regions were imaged at different times over the three Martian days to acquire consistent lighting and shadow conditions for all areas of the panorama. At left is a lander petal and a metallic mast, which is a portion of the low-gain antenna. On the horizon the double "Twin Peaks" are visible, about 0.6 to 1.2 miles away. The rock "Couch" is the dark, curved rock to the right of Twin Peaks. Another lander petal is at left center, showing the fully deployed forward ramp at the far left and the rear ramp at the right, which rover Sojourner used to descend to the surface of Mars on 5 July 1997.

included a variety of instruments, some of them U.S. equipment. A Japanese spacecraft, Nozomi, was set to arrive in orbit the same year to study solar winds and take photographs. In a historic mission to bring Martian soil and rock samples back to Earth for intensive study, NASA and the French space agency CNES planned to dispatch a robotic lander to the Martian surface. The agenda called for the rover to collect samples from diverse areas, place them in canisters aboard an ascent vehicle, and launch into Martian orbit. Meanwhile, an Ariane 5 booster from Khouru (the CNES launch site in French Guiana) had an appointment to blast off in 2005, carrying an additional lander to conduct soil tests. But the orbiter also had instructions for a maneuver to rendezvous with the canister-loaded vehicle, link up automatically, then shoot itself out of Martian orbit and return to Earth. With the data gleaned by these and other projects, scientists and engineers expected to mount an international mission to carry the first humans to a Martian landing sometime later in the twenty-first century.

Along with other members of an international research team, NASA continued its intensive investigation of Mars, where various phenomena seemed similar to those of Earth. By analyzing these similarities, scientists expected to learn much more about both planets. Some of the most striking news came from the Mars Global Surveyor, launched in 1997; mission specialists continued to review its scientific returns through the late 1990s and into the new century. The Surveyor's biggest piece of information was that Mars, like Earth, had experienced the physical effects of plate tectonics. Until the arrival of Mars Global Surveyor, the accepted view of Martian evolution rested on the idea that its crust consisted of a single shell, rather than interacting plates. The single crust theory put Mars into the same category as Mercury and the Moon—cold, stony, and lifeless. But the Surveyor's findings suggested a much more complex history for Mars—far more similar to that of Earth than previously believed. Because Earth's tectonic plate experience had such an impact on the evolution of advanced life forms, the possibility of some sort of parallel effects on Mars could not be ignored. On Mars, however, the process came to its finish over 4 billion years ago. The tantalizing new information yielded by the orbiting Mars Global Surveyor raised expectations for a next-generation Mars rover, known as FIDO (field integrated design and operations). During 1999, while undergoing tests in the California desert, its designers hailed FIDO as a device to eventually answer the question of life on Mars. FIDO's handlers looked forward to a historic mission in the twenty-first century.

Ongoing Research: From Earth to Galactic Space

NASA's investigations continued to embody a striking diversity. One project might roam through distant galaxies; another might cling close to home, scanning Earth's lower atmosphere; a third might sail off to study a corner of our own solar system.

One of the newest of these quests clearly looked to the future and to the prospects of encountering life forms elsewhere in the universe. In 1998 the agency funded an Astrobiology Institute with an annual budget of $10 million. A year later the institute's first permanent director was announced—Baruch Blumberg, a leading figure in medical anthropology, who received the Nobel Prize in 1976 for his work in medicine. Blumberg organized a compact staff at the institute's headquarters at NASA's Ames Research Center in Mountain View, California. The location provided a focused setting for specialists in biology, chemistry, physics, computing, and other studies. Administratively, the institute combined the expertise of eleven member organizations, ranging from Harvard's Woods Hole (Massachusetts) Marine Biological Laboratory to the Scripps Research Institute in California, the Johnson Space Center, and the Carnegie Institution. Its job: to continue during the twenty-first century the search for life in the universe.

Closer to home, NASA continued to analyze the dynamics of global warming, sometimes using some notably imaginative methods. One team of researchers pondered the role of high-altitude cirrus clouds in reflecting solar energy as a way to understand climatic effects and a possible influence on global warming. A related question involved "warm long wave radiation," a phenomenon in which the radiation—absorbed by clouds in the atmosphere—kept heat entrapped in the earth environment. To carry out observations, NASA worked with Sandia (New Mexico) Laboratories and conducted operations in the vicinity of Kauai, Hawaii. The project involved two aircraft with separate but integrated missions. One utilized a conventional DHC-6 Twin Otter, piloted by project personnel. The other featured an unmanned high-altitude aerial vehicle (UAV), known as the Altus. Both had radiometers and other instruments to measure solar energy and cloud characteristics such as optical properties and the nature of ice particles in clouds. Using data-linked Global Positioning System (GPS) signals beamed from the UAV, the Twin Otter aligned itself under the Altus as the two flew parallel paths to plot data with high precision. The project added valuable knowledge about natural as well as human-caused phenomena in the effort to understand global warming.

In several ways, the Cassini/Huygens mission to Saturn marked the close of one era of space exploration and the inauguration of a new one. Moreover, as space journalist Craig Couvalt wrote, the mission promised "to be one of the greatest explorations of the early 21st Century." Before Cassini, planetary landings on the Moon, on Mars, and on Venus resulted from ventures undertaken by the USSR or the United States alone. The new Saturn probe marked the first planetary exploration planned and organized by the international community. One of the project scientists described the huge planet and its myriad rings, 1 billion miles from Earth, as "the icon of interplanetary spaceflight." NASA's associate administrator for space science, Wesley Huntress, declared that "Cassini is clearly the most ambitious and complex deep space mission ever launched." As it lifted off in October 1997, Cassini had a planned trajectory to carry it into orbit around Saturn during the summer of 2004. At a cost of $3.3 billion, and weighing 12,600 pounds, Cassini became the costliest, heaviest planetary spacecraft project mounted by NASA in the twentieth century. Some observers saw Cassini as "a dying breed of large, expensive, multisensor spacecraft." At the same time, other factors influenced the shift to a new generation of smaller spacecraft with a more focused research design. In early phases of exploration to targets like Saturn—ten times the diameter of Earth, with eighteen moons and a system of rings spanning about 240,000 miles—it took large spacecraft with multiple capabilities to acquire any reasonable grasp the planet's complex dynamics. The next round of more compact planetary probes would concentrate on selected features.

Cassini's initial planning began in 1982, and a three-hundred-member international space team went to work at launch time to begin final plans for collection, dissemination, and analysis of data. Tasks for the Cassini segment included data gathering during seventy-four orbits of Saturn and elements of its rings and icy moons. The launch vehicle also carried the Huygens probe, the European Space Agency's first planetary lander, programmed to touch down on Titan, one of Saturn's moons. Aerospatiale acted as prime contractor for Huygens; DASA, the German space division of Daimler-Chrysler, handled test and integration tasks; ESA's member states included dozens of European subcontractors.

The U.S. Air Force supervised its launch from Cape Canaveral aboard an advanced Titan 4B with solid boosters in the lower stage and a Centaur liquid hydrogen upper stage. At the predawn liftoff from the Cape, the distant, winking light of Saturn itself hung above the towering launch vehicle. Assuming a successful mission, scientists expected Cassini to remain in orbit around Saturn for centuries, carrying a small, digital disc with the handwritten signatures of more than 616,000 people from eighty-one nations. This seemingly incidental gesture signified that the romance of space exploration still captured the imaginations of people everywhere—including the project manager, scientists, and engineers who signed off on the proposal to add this nonessential artifact to Cassini's framework.

In the meantime additional missions validated NASA's ongoing role in space exploration, reflected the agency's altered nature in terms of institutional culture and gender roles, underscored the legacy of international factors, and demonstrated increased interest in the dynamics of the global environment. In the summer of 1999, a second space telescope, the Chandra X-ray Observatory went into operation. Carried aboard *Columbia*, the launch marked another personnel milestone with NASA's first woman to command a shuttle mission, Eileen Collins. The telescope was named for the late Indian-American Nobel laureate Subramanyan Chandrasekhar. In contrast to the Hubble Space Telescope, which traveled around the world in a near-Earth orbit, the Chandra X-ray Observatory traveled in an oval-shaped path that ranged from 6,200 miles to 87,000 miles from Earth. During each 64-hour orbit, this track gave it 55 hours clear of radiation belts around the planet, permitting optimum environments to record X-ray images and similar data from ordinary comets to black holes and quasars at the edge of our universe. Scientific returns during its first two years proved to be highly rewarding. Another spacecraft, Terra, soared away from its launch pad at Vandenburg Air Force Base, California, on a mission to acquire masses of new data about Earth's land, oceans, air, ice caps, glaciers, forests, and human activities. As Terra began full-scale operations early in 2000, NASA scientists planned to consider the data collectively as a means to understand ways in which Earth's features interact within the global climate system.

One of the most appealing stories to result from a NASA space mission involved an asteroid called Eros, described as a peanut-shaped lump of rock approximately the area of Manhattan located 256 million miles from Earth. The spacecraft NEAR (Near Earth Asteroid Rendezvous probe) slipped into orbit about the asteroid on February 14, 2000—Valentine's Day. Given the asteroid's name, Eros, the coincidence provided plenty of playful copy for journalistic punsters. Using detailed infrared, X-ray, and gamma ray techniques, investigators expected to uncover considerable information about these fascinating space travelers, their origins, and their relationship to various galactic phenomena. After a few days, project officials realized that they had the opportunity to land the NEAR spacecraft on Eros itself, permitting unparalleled, close-up examination of the characteristics of an asteroid. Even though NEAR had not been engineered for such a maneuver, programmers patched up a series of coded commands; NEAR coasted up to Eros and gently settled on its surface. Elated scientists found that NEAR's instruments still worked, allowing analysis of the Eros surface from a distance of mere inches, rather than miles. Researchers cheerfully reported that the process of assessing NEAR's returns would keep them occupied into the coming century.

Aeronautics and Astronautics for the Twenty-First Century

Within the American aerospace community, multinational projects continued to develop; NASA maintained a high level of negotiations to develop collaborative efforts in the future as well as to maintain and enhance programs already under way. At the same time, many leaders in the aerospace community grew increasingly bothered by the dramatic advances achieved by European aerospace firms. Canadian, Asian, and Latin American manufacturers also became keen competitors in selected areas. Across the board—in space technology, civil aircraft, and military aeronautics—the American position in essential export sales came under relentless pressure from foreign suppliers. A decade earlier, international competition prompted NASA to redouble its efforts in selected areas of R&D. In 1997, to assist American companies in the global aerospace marketplace, NASA proclaimed a new broad-based initiative dubbed the Three Pillars. In general terms, NASA pledged to "stretch the boundaries of the knowledge and capabilities needed to keep the United States as the global leader in aeronautics and space." The agency also defined a three-part focus to emphasize astronautics, civil aviation, and air transportation safety in its Three Pillars programs. Several earlier programs had clearly foreshadowed the new enterprise. Elements of the Three Pillars could be discerned in nearly all of the subsequently announced programs, but civil aviation and air transportation system safety had clearly identifiable agendas in terms of meeting foreign competition. And although it was not officially cited in the Three Pillars, nobody at NASA expected work in military aviation to lag, either.

At Dryden, on just about any day of the week, visitors might spot one or more unusual aircraft taking off for an experimental flight. Dryden still flew two SR-71 aircraft on ultra-high-speed high altitude missions, plus a varying number of planes like the F/A-18 and F-15 fighter aircraft, usually with extensive modifications to wring out some new aerodynamic conundrum. Dryden also provided the home base for the Boeing 747 modified to transport the Space Shuttle, along with an elderly B-52, still used as a drop plane for sundry tests of aircraft designs in scaled-down form. Occasionally, other planes were based at Dryden, such as a Lockheed L-1011 TriStar and a McDonnell Douglas MD-11. In addition, Lockheed used Dryden's flight research facilities for flight testing the remotely piloted DarkStar surveillance aircraft. Despite the firm confidence of its test team, the DarkStar stalled, crashed, and burned, leaving a long black smudge beside the runway. Some mordant observers immediately referred to the site as DarkSpot. Lockheed went back to work on an improved model.

A lot of flying clearly implied military applications. One highly modified F/A-18, known as the Systems Research Aircraft (SRA), featured a digital fly-by-wire system subject to continuous alterations. A series of SRA flights included devices to shake the plane's wingtips in imitation of structural vibrations that could tear the plane apart. How could the pilot restore control? Were certain flight altitudes more destructive than others? Data from the SRA answered such questions. Then there was the F-15 ACTIVE, for Advanced Control Technology for Integrated Vehicles. On loan to Dryden from the air force, the ACTIVE plane had canards (actually a pair of F-18 stabilators) attached to the shoulders of its air intakes. Along with radically modified controls, the plane also featured swiveling exhaust nozzles on its powerful engines, continuing NASA's interest in vectored-thrust investigations. Among a series of flight experiments, one set of tests probed the causes for engine compressor stall at

A seven-year journey to the ringed planet Saturn began in 1999 with the liftoff of a Titan IVB/Centaur carrying the Cassini orbiter and its attached Huygens probe. After a 2.2-billion-mile journey requiring two swing-bys of Venus and one of Earth to gain additional velocity, the two-story-tall spacecraft was scheduled to arrive at Saturn in July 2004. The orbiter and its instruments were programmed for a four-year tour of duty around the planet.

high power settings. As usual, there was a stable of "X"-designated designs. The X-36, a scaled-down, remotely piloted aircraft, had special canard surfaces, a sharply indented wing along the line of its trailing edge, and no tail. It explored the nature of special designs to enhance stealth characteristics and verified control responses from its unusual canard/wing layout.

The presence of wide-body air transports clearly underscored NASA's commitment to America's airliner manufacturers. While a Lockheed L-1011 TriStar flew out of Dryden, a persistent research group studied careful adjustment of control surfaces within particular flight regimes in order to promote fuel efficiency. Such techniques could save hundreds of millions of dollars per year on global airline routes. In a different context, the MD-11 airliner carried the designation PCA, for Propulsion-Controlled Aircraft. In case of a disastrous failure in an airliner's flight control system, could it be landed by using only engine thrust for control? The NASA investigating team created a software package for crippled planes and carried out six successful landings without using the normal flight control system. All of these programs drew upon related research at Lewis, Langley, and Ames.

Over several decades, NACA had effectively assisted the U.S. aviation industry. By the mid-1990s, American manufacturers of air transports still commanded two-thirds of the international market. But the U.S. share of the market had eroded dramatically since 1985 owing to competition from Europe's Airbus Industrie. The demise of McDonnell Douglas owed much to the determination of Airbus, which hoped to capture about half of the airline market early in the twenty-first century. With the demand for subsonic aircraft estimated at $1 trillion between 1995 and 2015, further erosion of the American share would make a considerable dent in the national economy. Accordingly, NASA stepped up work in several areas of development, all designed to maintain market share as well as to enhance productivity and safety of the nation's air transport system. Several NASA centers provided significant assistance in subsonic air transport R&D. For example, after its venerable 737 went into retirement, Langley acquired a Boeing 757 as a Transport Systems Research Vehicle. NASA engineers modified it in order to conduct flight testing of advanced navigation and landing systems, flight controls, and weather aids. They also gave special attention to improved terminal area productivity, such as enhancing equipment and procedures for nonvisual operations during nighttime and bad weather.

There were several other areas under investigation. Because operating costs of jetliners often caused airlines to fly planes longer, NASA studied improved techniques for nondestructive testing to provide better safety evaluations for detecting corrosion, fatigue cracks, and so on. Environmental research focused on techniques to reduce harmful engine emissions and noise. Better use of composite materials in major flight components would reduce airframe weight, thereby cutting flying costs. Similarly, lightweight fiber optical systems would subtract pounds and provide better reliability. NASA also continued to study tilt-rotor aircraft, hoping to speed up their transition from the military inventory to effective use in the civil sector.

Because some investigations in these areas had been part of NASA's agenda for several years, new planes like the Boeing 777 realized early benefits from NASA work. Several centers contributed, although Langley played a principal role. Boeing engineers used a number of Langley's math procedures for computer-generated airflow images, and Langley's wind tunnels helped verify wing and engine nacelle installation. Other activities focused on noise reduction for passengers aboard and personnel in the terminal areas, tests of landing gear and tires, and better utilization of aerospace composite structures. Data from the Marshall Space Flight Center (MSFC), drawn from rocket engine turbine-wake patterns, contributed to jet engine improvements. In 1996, during a publicity tour in the first flight test version of the 777, Boeing pilots landed at Langley's airstrip to acknowledge NASA's participation in the plane's evolution.

Other tests of airliners used foreign equipment and occurred overseas. In the wake of the cold war, there were many examples of startling partnerships involving Russia and the United States. One of the most fascinating collaborations occurred when NASA used a refurbished TU-144 supersonic transport (SST) for specialized aerodynamic research. The Tupolev SST flew in 1968 and entered service in December 1976, following the first commercial flight of the Anglo-French Concorde SST in January of the same year. Eventually, the Soviet regime decided against subsidizing the TU-144 transports because of their high operating costs and other factors; the plane used by NASA was a TU-144LL, a later version built in 1981 for testing.

During the 1980s and 1990s, when NASA began to take another look at large civil transports in the SST configuration, engineers bemoaned the lack of a full-size airplane, since military designs like the ill-fated XB-70 were no longer in production. Eventually, talk turned to the Russian SST planes known to have survived. Using a TU-144 would allow engineers to compare actual, full-scale flight data against results from wind tunnel tests and computer simulations. By calibrating carefully planned test results, wind tunnel models and computer programs could be engineered to produce accurate data of all sorts. Such refined test results would save immense amounts of time and money in the design and construction of next-generation SST aircraft. After the cold war thaw,

NASA's query to Russian authorities eventually resulted in a deal. A modified, heavily instrumented TU-144LL acquired a NASA logo, with research flights in Russia beginning in 1996. Flown by both Russian and American pilots, the TU-144 completed nineteen research flights by 1998; both countries expressed satisfaction with the quality of the test data.

There were hybrid flight programs that involved exotic new designs, continuing the cutting-edge research that was the heritage of NACA's and NASA's X-series. Their missions, like the shuttle's, took in both aeronautics and astronautics. Two such programs utilized the X-33 and the X-38 technology demonstrators.

The X-38, with its stubby wings, owed some of its design to ideas borrowed from the lifting body series of small, experimental aircraft tested by NASA and the air force during the early 1970s. Its mission was to serve as an aerospace lifeboat, capable of safely returning up to seven crew members in case of a life-threatening emergency affecting the space station in orbit. The X-38 project was NASA's first human spacecraft design in over two decades. The 28.5-foot-long aircraft featured a carefully sculpted underbody—designed to develop a certain amount of lift—to assist its wings during reentry into Earth's atmosphere. Closer to landing, the X-38 deployed a large, rectangular parachute above its fuselage. Called a parafoil, its job was to aid in reducing speed and to provide more effective glide characteristics as the X-38 steered toward a touchdown. NASA reported good results following unmanned drop tests early in 1998; further tryouts were planned before two operational versions would be carried aloft aboard the shuttle and docked at the space station, probably in 2003.

The X-33 had a far more complex R&D history. Planned as a reusable rocket, the wedge-shaped vehicle employed lifting body technology along with stubby wings and vertical fins for added lift and control during liftoff or reentry. The X-33 program began during 1996, when the Clinton administration pressed NASA for a space transportation system that would be less expensive to operate than the Space Shuttle. In addition, the mission profile specified a single stage to orbit. Unlike the Space Shuttle, which shed solid rocket boosters and its huge propellant tank, the X-33 design emerged as a completely self-contained system. The contract for an initial X-33 flight article went to an industry team headed by Lockheed Martin, which had been conducting advanced studies on a similar spacecraft called VentureStar, carrying on a Lockheed tradition of celestial monikers for aerial craft (as in the Constellation of the 1940s–1950s propeller era; the TriStar jet airliner of the 1960s; and the F-80 Shooting Star fighter series, for example).

Although initial plans called for shuttle replacements in 2012, requiring test

A NASA SR-71 successfully completed its flight during October 1997 as part of the NASA–Rocketdyne–Lockheed Martin Linear Aerospike SR-71 Experiment *(LASRE)* at NASA's Dryden Flight Research Center. NASA intended the aerospike engine for use aboard the X-33 reuseable launch vehicle; the test module mounted on the SR-71 flew at speeds of Mach 1.2 in order to validate computational predictive tools.

flights of a prototype as early as 2000, NASA and industry experts alike recognized the daunting challenges of the X-33. As one NASA official remarked, the X-33 gave the government its most formidable space launch program in several decades. Most of the X-33's structure, thermal insulation, propellant tanks, and propulsion systems reflected major advances in the state of the art. In addition to special designs for the liquid hydrogen and oxygen tanks, plus many other systems, the X-33's liftoff power came from engines based on a linear aerospike concept, described as "a nozzleless propulsion system."

Happily, NASA continued to engage in flying unique—but weird-looking—aircraft that seemed to have little or no immediate application. These often began as shoestring projects cooked up by some peripheral organization that caught NASA's attention. Such was the Pathfinder project. Once joined with NASA, Pathfinder operated under its own weird acronym, ERAST, shorthand for Environmental Research Aircraft and Sensor Technology. In this case NASA acted as coordinator for a group of private firms that supplied labor and materials to design and build the aircraft. The machine was basically a unique, ninety-eight-foot-long flying wing covered with transparent mylar film and powered by six electric motors that derived their energy from solar cells incorporated into the wing structure. But the plane needed a mission.

Many of NASA's research efforts concerned Earth-related programs such as environmental studies, oceanography, and agricultural production. During the late 1990s, NASA scientists devoted special attention to the continuing debate about ozone holes and general ozone depletion in Earth's atmosphere. Using various climate models, agency studies pointed to "greenhouse effect" gases as a contributing factor. Addressing the need for inexpensive but long-duration atmospheric flights to gather significant data, NASA kept the Pathfinder project alive at ERAST. By 1998 Pathfinder-Plus (with a 121-foot wingspan) set a new altitude record of over 80,000 feet, the highest ever achieved by a propeller-driven aircraft. In the meantime other research efforts began to refine procedures that Pathfinder's progeny might use in gathering atmospheric research data. Iconoclastic designers were already at work on a successor named Helios, a creation with a wingspan of 250 feet (longer than the wings of a Boeing 747) and powered by fourteen solar-cell electric motors each having about the strength of a hair dryer motor. Even though it could only cruise at about the speed of a bicycle, its builders planned to send Helios climbing to 100,000 feet. In August 2001 Helios clawed its way to 96,863 feet—a convincing demonstration cut short by diminishing sunlight. The mission convinced researchers that improved solar-cell technology would give Helios a good operational altitude of 50,000 feet, allowing it to store enough daytime solar energy to cruise through the night and remain aloft for weeks at a time while gathering scientific information.

As usual, diversity characterized NASA's stable of test aircraft and other aviation assets. With hardware and test items sent between centers and contractors, the increasing tempo of ISS activities put a premium on timely deliveries. As in the days of the Apollo program, the large dimensions of many ISS components argued against truck or rail shipment. Again, NASA turned to air transport to solve the problem—an updated version of the Super Guppy used during the 1960s for the Apollo program. Turboprop variants of the plane were flown by NASA through 1991. NASA's new Super Guppy, delivered in the autumn of 1997, had been expressly modified for Airbus Industrie in Europe. The plane carried wing sections, fuselage segments, and other oversized components of Airbus jet airliners from several European countries to aircraft assembly locations in France and Germany. Built from surplus Boeing C-97 transports, the enlarged Super Guppy fuselage could take on pieces of equipment too large even for the U.S. Air Force's Lockheed C-5 transport. Loading was through the plane's hinged nose section, and the Super Guppy's cargo area measured 25 feet high by 25 feet wide by 111 feet long. Budget-minded NASA managers bartered for the unique plane through a deal worked out with the Europeans, offering research space on two upcoming shuttle flights in return for the Super Guppy. Meanwhile, it carried out essential tasks in its new assignments with NASA.

In the process of closing out the decade, the presence of some twenty thousand NASA employees indicated the nation's commitment to continuing the progress of women and minority groups in securing satisfying vocations and achieving leadership positions. The agency took pride in operating realistically despite pinched budgets. NASA, proclaimed Administrator Goldin, had set the example of a federal bureaucracy that did its job effectively, within budget and still carried on an R&D program for flight and space exploration that was based on startling expectations. During 1999, when the agency still encountered stubborn congressional hostility over its $13.6 billion for the next budgetary cycle, NASA officials and aerospace leaders waged a belligerent campaign to get funding restored. Among other things, they cited the need for the American aerospace community to continue broad-based research to meet international competition, and they pointed out that NASA stood out as the American champion in this global jousting for commercial advantage. Such overt public and political maneuvering did not reflect historical NACA habits. But for NASA in the post–cold war era, bureaucratic skirmishes like this became standard operating procedure and seemed likely to continue.

Chapter 11

Retrospect and Prospect

During the halcyon era between World War I and World War II, NACA's work on airfoils, engine cowlings, icing, and other problems drew the attention of aeronautical engineers around the world. There were also institutional changes, especially in the 1930s, when the agency became more attuned to industry trends and more politically aware in its interaction with congressional committees. World War II brought the more dramatic shifts: research intimately geared to national security, growth from one small facility to three elaborate centers sited coast to coast, and ballooning budgets and personnel rosters. For all its successes, the agency also lost some of its luster as European advances in gas turbines and high-speed flight received postwar attention.

The postwar era entailed cold war tensions and national security budgets that promoted advanced flight research. NACA flourished. Cooperative programs with the military brought the X-1 and X-15 into being. These programs also added responsibilities for design and program management to NACA's tradition of research and flight testing. The old "advisory" committee became a large bureaucracy for R&D.

The shock of the successful Soviet launch of *Sputnik* in 1957 altered NACA forever. Granted billion-dollar budgets by Congress, the new NASA was thrust into an international spotlight as America's answer to the Soviet Union for leadership in space exploration. With four new centers, NASA rapidly developed skills in the novel field of astronautics. Personnel also had to build new skills to manage huge budgets and mature aerospace contractors scattered across the continent. The spotlight of the space race also intensified the agency's problems when projects missed deadlines and when astronauts died. Nonetheless, Apollo was a successful effort and a historic achievement. Although issues of American and Soviet competition for global influence colored the origins of the program and the triumphant voyage of *Apollo 11,* the new awareness of the fragile existence of Earth within our universe also fostered a promising spirit of international cooperation.

NASA's mission and purpose were not always clear in the post-Apollo era. The sense of urgency that spurred Apollo had dissipated. In aeronautics NASA made sure progress in hypersonic flight and began highly beneficial programs to control pollution, reduce engine noise, and enhance fuel economy—programs that assumed growing importance in an environmentally conscious society. In astronautics the Space Shuttle became a fascinating operational program, although critics maintained that it was a complex system with no major or scientific mission to justify its expense. A proposed space station, which would absorb numerous shuttle flights, was persistently plagued by budget issues. The loss of *Challenger* in 1986 underscored the risk of relying so heavily on the shuttle at the expense of expendable launch vehicles. Reorganizing priorities for military and civil payloads proved to be a frustrating exercise. A renewed wave of criticism concerning lower budgets for space science surfaced, a reminder of controversies over manned versus unmanned flights that had been going on since the early days of the space program. There was also concern stemming from various studies that noted the constraining effects that seemed endemic to large bureaucracies, as well as the demographic realities of a workforce—heavily recruited in the 1960s—that might lose its sense of adventure as the time for retirement loomed.

For NASA the decade of the 1990s brought about an encouraging transition. New management approaches helped restore an adventurous momentum, and a cadre of younger personnel joined the agency. They were motivated by opportunities to explore new regimes in flight research, fascinated by the scope of the ISS, and challenged by complex missions to outer space. NASA also paid more attention to spin-off programs and other initiatives keyed to issues that might engage broader congressional support. The widely publicized Space Shuttle mission in 1998 that included seventy-seven-year-old Senator John Glenn prompted criticism from many quarters but also supplied useful medical records, including baseline data applicable to older manned spaceflight personnel on future missions. The facts that Glenn's reprise of his 1962 orbital flight pleased innumerable space partisans and delighted thousands of the world's senior citizens—and attracted global news coverage—made it all seem worthwhile. During 1999 NASA decided it would be appropriate to revise the name of the Lewis Center at Cleveland, Ohio (Glenn's home state), to John H. Glenn Research Center.

Nonetheless, on entering the twenty-first century, NASA found itself buffeted by the winds of change. In its initiatives to support a broad spectrum of projects for applied science and technology in daily life, NASA ventured far from its aeronautical origins in 1915. Budgetary gaps in the ISS forced program cuts elsewhere, including foreclosure on several high-profile projects. The X-33, beset by major troubles with its propellant storage systems, became a victim, along with the X-34 technology test-bed program to investigate reusable space vehicles. Even shuttle-related programs like the X-38 emergency vehicle faced a problematic budgetary future. Struggles to cope with decreased resources continued. Within the decade of the 1990s, NASA's civil service roster declined from 24,000 to 18,000, a trend that often required stiffer workloads for key personnel. Although the agency expected a budget of approximately $14.5 billion for fiscal year 2002, that amount represented a decline of some $150 million in real dollars when compared to funding in recent years. Critics in the U.S. aerospace community as well as in Congress worried that the agency had allowed American leadership in aeronautics to slip. These rumblings were anything but haphazard, emerging from such respected professional organizations as the American Society for Mechanical Engineers, the American Institute of Aeronautics and Astronautics, and the editors of leading periodicals in the field, *Aviation Week and Space Technology,* for example. Other sources claimed that elaborate manned ventures, specifically the International Space Station, siphoned off funds that should be used for aeronautics as well as space science. Broader budgetary and technological collaboration with foreign partners also underscored new variables in NASA programs.

In November 2001, after nine and one-half years in office, Daniel Goldin departed as the NASA administrator. During his term the agency launched 171 space missions and recorded 160 of them as successful. Overall, this was an impressive record of technical prowess, achieved while NASA trimmed its civil service roll by 25 percent and cut its contractor workforce by 10,000 personnel. It fell to Goldin's successor to restore momentum in aeronautical programs and to sort out formidable budgetary conundrums. The new administrator, Sean O'Keefe, had been deputy director of the Office of Management and Budget and a former secretary of the navy.

Aside from these administrative issues, the responsibilities of R&D continued. At Dryden officials announced a new program, RevCon, intended to accelerate the development of high-risk, breakthrough technologies in the design of aeronautical craft. Elsewhere, at a press conference in the spring of 2001, NASA representatives glumly reported on the first test flight of an unmanned X-43A test vehicle with an exotic hydrogen engine design, known as a "scramjet." A failure in the plane's control system had led to its demise. Stoic engineers began reviewing flight data and engineering drawings in preparation for a new test. In a different sort of test, a research team successfully "landed" a simulated 757 transport using nerve signals transmitted through movements of the pilot's forearm. The process represented a major step forward in neuro-electric control procedures, relying on electrodes implanted in an armband to provide "rapid, intuitive control." Proposed applications ran the gamut, from aerial combat, to microsurgery, to manipulation of hard-to-use tools by astronauts in space. At Cape Canaveral another press conference highlighted the launch of the Microwave Anisotropy Probe (MAP), on a million-mile trip to its designated position in space. Late in 2002 project scientists expected MAP to begin returning data about microwaves that began their journeys some four hundred thousand years ago, answering basic questions about the "big bang" theory and the origins of the universe.

The problems of flight, whether aircraft or spacecraft, still pervaded the agency's principal activities and continued to motivate thousands of civil service and contractor professionals devoted to the myriad challenges of aerospace research.

Notes on Further Reading

The quotations in this book can be found in the following publications. In chapter 1, the comment on Langley's variable density tunnel comes from James Hansen, *Engineer in Charge: A History of the Langley Aeronautical Laboratory, 1917–1958* (Washington, D.C.: U.S. Government Printing Office, 1987), 84; the characterization of Robert Goddard from Frank Winter, *Prelude to the Space Age: The Rocket Societies: 1924–1940* (Washington, D.C.: Smithsonian Institution Press, 1983), 21–22. In chapter 2, John Becker's description of early engineering experiences at Langley appears in his memoir, *The High-Speed Frontier: Case Histories of Four NACA Programs, 1920–1950* (Washington, D.C.: U.S. Government Printing Office, 1980), 22. Alex Roland's critique of NASA in chapter 3 comes from his *Model Research: The National Advisory Committee for Aeronautics, 1915–1958*, vol. 1 (Washington, D.C.: U.S. Government Printing Office, 1985), 194. In chapter 4, John Becker's remarks about the X-15 are quoted in Richard Hallion, *On the Frontier: Flight Research at Dryden, 1946–1981* (Washington, D.C.: U.S. Government Printing Office, 1984), 107; President Kennedy's comment about catching up with the Soviets in space is quoted in Lloyd S. Swenson Jr., James M. Grimwood, and Charles C. Alexander, *This New Ocean: A History of Project Mercury* (Washington, D.C.: U.S. Government Printing Office, 1966), 335; Kennedy's lunar landing speech to Congress is excerpted in John Logsdon, ed., *Exploring the Unknown: Selected Documents in the History of the U.S. Civil Space Program*, vol. 1 (Washington, D.C.: U.S. Government Printing Office), 453. From chapter 5, Werner von Braun's complaint about the heavy lunar module is quoted in Roger Bilstein, *Stages to Saturn: A Technological History of the Apollo/Saturn Launch Vehicles* (Washington, D.C.: U.S. Government Printing Office, 1980), 193. In chapter 8, the comparison between *Viking 2* and William Tell's arrow appeared in an article by Rick Gore, "Uranus: Voyager Visits a Dark Planet," *National Geographic,* August 1986, 182. The astrophysicist's quote in chapter 9 appeared in an article by Dick Thompson, "Big Gamble in Space," *Time,* March 22, 1993, 63. The material about the Cassini/Huygens mission in chapter 10 was reported by Craig Covault, "Saturn's Mysteries Beckon Cassini," *Aviation Week and Space Technology,* October 20, 1997, 22–24.

Background

Although somewhat dated, an aerospace bibliography prepared by the staff of the National Air and Space Museum not only provides an annotated, comprehensive guide to both American and international sources but also offers a fine review of other bibliographies: Dominick A. Pisano and Cathleen S. Lewis, eds., *Air and Space History: An Annotated Bibliography* (New York: Garland, 1988). The American chapter of early aeronautics is definitively recounted by Tom D. Crouch, *A Dream of Wings: Americans and the Airplane, 1875–1905* (New York: Norton, 1981). For a combined history of American aviation and space exploration, see Roger E. Bilstein, *Flight in America: From the Wrights to the Astronauts*, 3d ed. (Baltimore: Johns Hopkins University Press, 2001). Especially for its numerous illustrations, Wernher von Braun and Fred I. Ordway III, *History of Rocketry and Space Travel* (New York: Thomas Y. Crowell, 1975), is still a useful survey of astronautics. Roger D. Launius, *Frontiers of Space Exploration* (New York: Greenwood Press, 1998), covers the entire spectrum from early rocketry to American and Russian experiences with *Mir* and also assesses major unmanned missions such as the Mars probes of 1997. There is a good chronology along with skillfully chosen documents. A series of scholarly essays with special attention to American topics is included in Eugene Emme, ed., *The History of Rocket Technology: Essays on Research, Development, and Utility* (Detroit: Wayne State University Press, 1964). The Pulitzer Prize–winning study by Walter McDougal, *The Heavens and the Earth: A Political History of the Space Age* (New York: Basic Books, 1985), analyzes the American and Soviet space programs as part of the cold war and technocratic trends. For the definitive account of the Soviet effort, see Asif A. Siddiqi, *Challenge to Apollo: The Soviet Union and the Space Race, 1945–1974* (Washington, D.C.: Government Printing Office, 2000).

The NASA History Office has sponsored a series of monographs about specific NASA centers, as well as specialized studies on aspects of aviation and space. Although this survey rests heavily on these volumes, I have not cited all of them below; a complete list of NASA History Series titles is available from the NASA History Office, Code ZH, NASA Headquarters, Washington, D.C. 20546. The Web site is http://history.nasa.gov.

NACA and Aviation to 1958

NACA's origins, technical contributions, and political evolution have been thoroughly assessed by Roland, *Model Research,* in two volumes. The first volume is a historical narrative; volume 2 contains annotated documentation. Roland criticizes the politicization of the agency. Hansen, *Engineer in Charge,* covering the same time span, focuses on Langley's research functions. Virginia Dawson has skillfully written *Engines and Innovation: Lewis Laboratory and American Propulsion Technology* (Washington, D.C.: U.S. Government Printing Office, 1991), discussing piston engines and jet engines as well as space systems. The organization and early years of NACA's Ames facility are the subjects of Elizabeth A. Muenger, *Searching the Horizon: A History of Ames Research Center, 1940–1976* (Washington, D.C.: U.S. Government Printing Office, 1985).

General trends in the aerospace business can be traced in Roger E. Bilstein, *The Enterprise of Flight: The American Aviation and Aerospace Industry* (Washington, D.C.: Smithsonian Institution Press, 2001). For specific technical development by individuals and organizations besides NACA, see Ronald Miller and David Sawers, *The Technical Development of Modern Aviation* (New York: Praeger, 1970). The fascinating story of the jet engine and Europe's leadership in this field can be found in Edward W. Constant II, *The Origins of the Turbojet Revolution* (Baltimore: Johns Hopkins University Press, 1980). The monographs by Roland, Hansen, and Dawson, cited above, present other carefully argued viewpoints.

For an informative look at early rocket societies in America and abroad, see Winter, *Prelude to the Space Age*. On the background of German rocketry and Wernher von Braun, see the popularly written study by Frederick I. Ordway III and Mitchell R. Sharpe, *The Rocket Team* (New York: Thomas Y. Crowell, 1979), which is based on extensive interviews. In his prize-winning study, Michael J. Neufeld, *The Rocket and The Reich: Peenemuende and the Coming of the Ballistic Missile Era* (New York: Free Press, 1993), takes a more critical approach.

For a summary of NACA's early postwar aerodynamic activities, see Hansen, *Engineer in Charge*. The story of the X-1 and the early challenge of the "sonic barrier" are detailed in Richard P. Hallion, *Supersonic Flight: Breaking the Sound Barrier and Beyond* (New York: Macmillan, 1972). The story of Michael Gluhareff and the swept wing is recounted in an article by the same author, "Lippisch, Gluhareff, and Jones: The Emergence of the Delta Planform and the Origins of the Sweptwing in the United States," *Aerospace Historian* 26 (March 1979): 1–10.

NASA Origins and the Apollo Era

A series of NASA-sponsored histories covers the transition from NACA to the new NASA and the progress of the Apollo program. The background of the IGY and America's initial plans to launch a satellite is the subject of Constance Green and Milton Lomask, *Vanguard: A History* (Washington, D.C.: Smithsonian Institution Press, 1971). The drama of the NACA-NASA transition and the difficulties of launching a coherent space program are clearly set out in Swenson, Grimwood, and Alexander, *This New Ocean*. The interaction of space science and cold war agendas associated with the "national security state" are dissected by David H. De Vorkin, *Science with a Vengeance: How the Military Created the U.S. Space Sciences after World II* (New York: Springer-Verlag, 1992).

As attention began to focus on manned lunar missions, politics and technology became increasingly entwined; this story is told by John M. Logsdon, *The Decision* to Go to the Moon: Project Apollo and the National Interest (Cambridge, Mass.: MIT Press, 1970). Manned launches unquestionably provided drama during the space missions of the 1960s. The Mercury program is covered by Swenson, Grimwood, and Alexander in *This New Ocean*. For the official history of the next manned phase, see Barton C. Hacker and James M. Grimwood, *On the Shoulders of Titans: A History of Project Gemini* (Washington, D.C.: U.S. Government Printing Office, 1977). The Apollo missions (through Apollo 11), which formed the centerpiece of America's manned space effort during the decade, are the subject of Courtney G. Brooks, James M. Grimwood, and Lloyd S. Swenson Jr., *Chariots for Apollo: A History of Manned Lunar Spacecraft* (Washington, D.C.: U.S. Government Printing Office, 1979). The success of Apollo required development of a family of large launch vehicles and a sophisticated launch complex. These topics are covered in Bilstein, *Stages to Saturn*; and Charles D. Benson and William B. Faherty, *Moonport: A History of Apollo Launch Facilities and Operations* (Washington, D.C.: U.S. Government Printing Office, 1978).

Although launches from Cape Canaveral inevitably drew hundreds of thousands of enthusiastic spectators, public support of the space program was far from unanimous. A number of writers criticized the program as a cynical mix of public relations and profit-seeking, a massive drain of tax funds away from serious domestic ills of the decade, or a technological high card in international tensions during the cold war. See, for example, Edwin Diamond, *The Rise and Fall of the Space Age* (Garden City, N.Y.: Doubleday, 1964); Amitai Etzioni, *The Moondoggle: Domestic and International Implications of the Space Race* (Garden City, N.Y.: Doubleday, 1964); Vernon van Dyke, *Pride and Power: The Rationale of the Space Program* (Urbana: University of Illinois Press, 1964).

Richard S. Lewis, a highly regarded scientific journalist of the era, has written a balanced assessment, *The Voyages of Apollo: The Exploration of the Moon* (New York: Quadrangle, 1974). Tom Wolfe, *The Right Stuff* (New York: Farrar, Straus & Giroux, 1979), is a scintillating essay that emphasizes the personalities of the astronauts. Although astronauts are not necessarily skillful authors, Michael Collins, *Carrying the Fire: An Astronaut's Journeys* (New York: Farrar, Straus & Giroux, 1974), is an exceptionally well-written memoir that is notable for its lucidity as well as its modesty. For insights about the key role played by ground controllers, Gene Kranz's autobiography is especially interesting because it is backed by extensive notes that he kept during his career. See *Failure Is Not an Option: Mission Control from Mercury to Apollo 13 and Beyond* (New York: Simon & Schuster, 2000). James Harford, former director of the American Institute of Aeronautics and Astronautics, has written an incisive biography, *Korolev: How One Man Masterminded the Soviet Drive to Beat America to the Moon* (New York: John Wiley & Sons, 1997), on the principal leader of the Soviet program. William E. Burrows, *This New Ocean: The Story of the First Space Age* (New York: Random House, 1998), is a comprehensive survey of both American and Soviet/Russian efforts that colorfully covers highlights and also includes commentary on political angles and leading personalities.

The Post-Apollo Years

A trio of NASA-sponsored monographs deals with the principal programs of the early post-Apollo era. Edward C. Ezell and Linda Neuman Ezell, *The Partnership: A History of the Apollo-Soyuz Test Project* (Washington, D.C.: U.S. Government Printing Office, 1978), is a fascinating record of the negotiations and technical adjustments necessary to bring American and Soviet manned spacecraft together in orbit. There had been considerable criticism of NASA's emphasis on manned missions, a bias that many observers felt had hindered progress in space science. This issue was somewhat ameliorated by the spectacular unmanned Mars probes of the late 1970s. The Ezell writing team detailed these activities in *On Mars: Exploration of the Red Planet, 1958–1978* (Washington, D.C.: U.S. Government Printing Office, 1984). There was also a significant volume of space science undertaken in the manned missions of Skylab, carefully and skillfully explained by W. David Compton and Charles D. Benson, *Living and Working in Space: A History of Skylab* (Washington, D.C.: U.S. Government Printing Office, 1983).

Science is also an important theme in Clayton R. Koppes, *JPL and the American Space Program* (New Haven, Conn.: Yale University Press, 1982), a book that also elucidates the relationships between NASA and its contractors, including the academic community. Space science is the principal theme of Homer E. Newell, *Beyond the Atmosphere: Early Years of Space Science* (Washington, D.C.: National Aeronautics and Space Administration, 1980). Because Newell was a central figure during the years of Vanguard through the shuttle plans of the early 1970s, his book is a valuable memoir. For a reflective survey, see Paul A. Hanle and V. Chamberlin, eds., *Space Science Comes of Age: Perspectives in the History of the Space Sciences* (Washington, D.C.: Smithsonian Institution Press, 1982).

Elizabeth A. Muenger, *Searching the Horizon: A History of Ames Research Center, 1940–1976* (Washington, D.C.: U.S. Government Printing Office, 1985), discusses this center's important role in aeronautics as well as astronautics. James R. Hansen, *Spaceflight Revolution: NASA Langley Research Center from Sputnik to Apollo* (Washington, D.C.: U.S. Government Printing Office, 1995), similarly analyzes Langley's efforts to balance aviation and space ventures. Henry C. Dethloff, *"Suddenly Tomorrow Came . . . ": A History of the Johnson Space Center* (Washington, D.C.: Government Printing Office, 1993), is an incisive survey of a major center in transition during the post-Apollo era.

NASA's continuing work in high-speed flight research is chronicled by Hallion, *On the Frontier*, a book that covers the X-15, lifting bodies, and the evolution of the Space Shuttle. Lane Wallace, *Flights of Discovery: 50 Years at the NASA Dryden Flight*

Research Center (Washington, D.C.: Government Printing Office, 1996), is a handsomely illustrated publication that stresses technological milestones. David A. Anderton, *Sixty Years of Aeronautical Research, 1917–1977* (Washington, D.C.: U.S. Government Printing Office, 1978), is a concise, well-illustrated summary. Although it focuses on Langley and offers little interpretation, it is a useful guide to NACA and NASA aviation programs. Arthur Pearcy, *Flying the Frontier: NACA and NASA Experimental Aircraft* (Annapolis: Naval Institute Press, 1993), combines illustrations and an informed narrative in a broad survey of many different airplanes.

Aerospace Initiatives

The NASA History Office has sponsored a number of projects on various aspects of the Space Shuttle, planetary probes, applications satellites, space science, the space station, university/contractor relations, cultural responses to flight, aerospace research and development, and other topics.

Hallion, *On the Frontier*, provides an informative survey of high-speed aeronautical experimentation as well as useful flight test information about the shuttle. Howard Allaway, *The Space Shuttle at Work*, NASA SP-432 (1980), a NASA brochure released on the eve of shuttle operational flights, provides good technical background and mission plans. For a more comprehensive survey, see T. A. Heppenheimer, *The Space Shuttle Decision: NASA's Search for a Reusable Space Shuttle* (Washington, D.C.: NASA History Office, 1999).

The loss of the space shuttle *Challenger* caused considerable introspection about NASA management and the process of research and development for space exploration. The event is officially assessed in *Report of the President's Commission on the Space Shuttle Challenger Accident* (Washington, D.C.: U.S. Government Printing Office, 1968), which offers insights into NASA's political, technical, and managerial characteristics. The agency became the target of many critical books and articles that not only dissected the *Challenger* incident but also discussed perceived flaws throughout the NASA structure. See, for example, Joseph J. Trento, *Prescription for Disaster: From the Glory of Apollo to the Betrayal of the Shuttle* (New York: Crown Publishers, 1987). Alex Roland, "The Shuttle: Triumph or Turkey?" *Discover* 6 (November 1985): 29–49, a cautionary assessment of the shuttle, appeared three months before *Challenger*'s last mission. For a scholarly assessment, see Diane Vaughan, *The Challenger Launch Decision: Risky Technology, Culture, and Deviance at NASA* (Chicago: University of Chicago Press, 1996). For an informed analysis of NASA's management environment over nearly two decades, consult Howard E. McCurdy, *Inside NASA: High Technology and Organizational Change in the U.S. Space Program* (Baltimore: Johns Hopkins University Press, 1993). Andrew J. Dunar and Stephen P. Waring, *Power to Explore: A History of Marshall Spaceflight Center, 1960–1990* (Washington, D.C.: Government Printing Office, 1999), not only provides a cogent dissection of the *Challenger* issue but also offers informative insights into the broad activities of a major NASA center.

A sense of NASA's varied efforts in energy research, aeronautics, and space science over the past several years can be found in *NASA the First 25 Years, 1958–1983,* NASA EP-182 (1983). NASA has released numerous brochures pertaining to specific projects and missions. See, for example, *Galileo to Jupiter: Probing the Planet and Its Moons,* Jet Propulsion Laboratory, JPL 400-15 (1979); and Joseph J. McRoberts, *Space Telescope,* NASA EP-166 (n.d.). These and a wide range of NASA news releases are well illustrated and useful sources. See also NASA's colorful and informative annual report, *Spinoff* (1976 to present), which summarizes aeronautical research and development as well as major space ventures. This publication also chronicles technologies that are being applied or are potentially useful in the marketplace.

NASA's growing involvement in international initiatives has focused more attention on European and Russian operations. For an authoritative, informative introduction to the European Space Agency, see John Krige and Arturo Russo, *Europe in Space, 1960–1973* (Noordwijk, Netherlands: European Space Agency, 1994). For a comprehensive overview, see Roger Bonnet and Vittorio Manno, *International Cooperation in Space: The Example of the European Space Agency* (Cambridge, Mass: Harvard University Press, 1994), which carries the story into the 1990s. For the Russian side of things, see Phillip Clark, *The Soviet Manned Space Program* (New York: Orion Books, 1988), by a British expert in astronautics. Popularly written, the book has a wealth of information about individual cosmonauts, along with early manned launches involving Soyuz and Salyut and an analysis of the origins of *Mir.* See also David S. F. Portree, *Mir Hardware Heritage* (Houston: NASA/Johnson Spaceflight Center, 1994). Loaded with illustrations, this is an invaluable aid for following Soviet and Russian crews and for understanding the evolution of Soviet stations, particularly the development and complexity of the *Mir* space station.

The evolution of Space Shuttle missions and their eventual role in early international space station preparations constitutes another complex story. Judy A. Rumerman, *U.S. Human Spaceflight: A Record of Achievement, 1961–1998* (Washington, D.C.: NASA/Headquarters History Division, 1998), presents an invaluable chronology that ranges from early Mercury launches up though the final shuttle linkups with the Russian *Mir.* John Logsdon, *Together in Orbit: The Origins of International Participation in the Space Station* (Washington, D.C.: NASA/Headquarters History Division, 1998), is a carefully detailed, monographic summary of discussions from the 1970s to 1988. Howard E. McCurdy, *The Space Station Decision: Incremental Politics and Technological Choice* (Baltimore: Johns Hopkins University Press, 1990), elaborates on NASA, congressional politics, and the diplomatic evolution of plans for an international space station. Russian and American crews aboard *Mir* required a melding of two different traditions of manned spaceflight.

This aspect is the focus of Bryan Burroughs, *Dragonfly: NASA and the Crisis aboard Mir* (New York: HarperCollins, 1998). Covering eleven residents of *Mir* during 1995–98, the book is written from an American viewpoint and emphasizes problems.

Several recent studies have analyzed NACA and NASA from different historical perspectives. Roger Launius and Howard E. McCurdy, eds., *Spaceflight and the Myth of Presidential Leadership* (Urbana: University of Illinois Press, 1997), considers the shifting political goals and roles of the executive office in its relationships with Congress and the vagaries of public opinion. Pamela E. Mack, ed., *From Engineering Science to Big Science: The NACA and NASA Collier Trophy Research Project Winners* (Washington, D.C.: U.S. Government Printing Office, 1998), covering both aviation and space topics, provides insights into the evolution of the agency's technical expertise and its methods of translating research into operational systems and hardware. Joan Lisa Bromberg, *NASA and the Space Industry* (Baltimore: Johns Hopkins University Press, 1999), reflects on the agency's relations with the private sector, with informative commentary on the expertise that the private sector brought to NASA's ventures. Popular responses to America's space program over several decades are dissected in Howard D. McCurdy, *Space and the American Imagination* (Washington, D.C.: Smithsonian Institution Press, 1997), which explores the cultural mythology of the frontier, faith in technological progress, heroic explorers, and the vagaries of space exploration.

For recent trends in aeronautics and astronautics, I have relied on many detailed articles in the authoritative periodical *Aviation Week and Space Technology;* the journal published by the American Institute of Aeronautics and Astronautics, *Aerospace America;* and its counterpart in Great Britain, *Aerospace International,* issued by the Royal Aeronautical Society.

NACA / NASA Chronology

1903	Orville Wright makes the first sustained, powered, and controlled flight in an airplane.
1913	First college-level courses in aeronautical engineering offered at University of Michigan and Massachusetts Institute of Technology.
1914	World War I begins in Europe.
1915	National Advisory Committee for Aeronautics (NACA) is founded, largely in response to impressive European leadership in aviation.
1917	United States enters World War I; until Armistice in 1918, NACA is active in recommending areas in military aeronautics needing analysis and research.
1920	Langley Research Center (Hampton, Va.) dedicated as NACA'S first aeronautical research facilities.
1921	Authorization given for NACA variable density wind tunnel.
1924	Central Committee for the Study of Rocket Propulsion established in the USSR.
1926	Robert Goddard successfully launches the first liquid propellant rocket.
	First annual NACA inspection and conference for industry and other aviation representatives hosted at Langley.
	Largest wind tunnel in world (twenty-foot throat) constructed at Langley.
1927	Charles Lindbergh completes first transatlantic nonstop solo flight.
	Society for Space Travel (VfR) formed in Germany.
1928	NACA low-drag cowling enables dramatic increases in speed for airplanes powered by radial air-cooled engines.
1929	Lieutenant James H. Doolittle makes first takeoff and landing using instruments only ("all-blind flight").

Collier Trophy, awarded by the U.S. president for the year's "greatest achievement in aviation in America," goes to NACA for its low-drag cowling.

1930   American Interplanetary Society founded in New York City (named American Rocket Society in 1934).

1931   First full-scale wind tunnel for testing complete airplanes dedicated at Langley.

1932   Institute of Aeronautical Sciences incorporated in New York. Becomes American Institute of Aeronautics and Astronautics in 1963 and includes American Rocket Society.

1933   NACA summary report on airfoil sections becomes "designer's bible" for future generations of aircraft.

1935   Robert Goddard launches first rocket with gyroscopic controls.

1936   Theodore von Karman, at California Institute of Technology, founds research group eventually known as Jet Propulsion Laboratory (JPL).

1937   First static test of a gas turbine engine by Frank Whittle (United Kingdom).

USSR establishes rocket test centers in Moscow, Leningrad, and Kazan.

1939   U.S. Congress authorizes construction of second NACA research station: Ames Aeronautical Laboratory (California). Renamed Ames Research Center in 1958.

Heinkel HE 178 (Germany) makes first successful flight by a jet airplane.

World War II begins in Europe.

1940   U.S. Congress authorizes third new research station: Aircraft Engine Research Laboratory (Ohio). Renamed Lewis Center in 1948, John H. Glenn Center in 1999.

1941   Pearl Harbor attacked by Japanese carrier aircraft; U.S. enters World War II.

1942   First test of Germany's V-2 rocket, developed by Wernher von Braun team, at Peenemünde (Germany).

1945   Project Paperclip organized by U.S. authorities to bring leading German aviation and rocket experts to work in America.

Beginning in the late 1930s and continuing through the end of World War II, NACA theoretical and applied research makes major contributions to military aircraft.

1946   NACA Muroc Flight Test Unit established (California); carries out work in advanced aeronautics and high-speed flight. After various designations, named Dryden Flight Research Center in 1994.

1947   Bell XS-1, developed by NACA in collaboration with the USAF, becomes first aircraft to exceed speed of sound.

        NACA and USAF partnership begins landmark series of pioneering X-aircraft in high-speed flight research.

1948   Organization for European Economic Cooperation established; significant step toward a unified Europe and cooperative aeronautical research and development.

1949   North Atlantic Treaty Organization; symbol of cold war antagonisms involving the USSR through the early 1990s.

        Unitary Wind Tunnel Program becomes policy.

1952   De Havilland Comet (Britain), world's first jet airliner, enters service.

1957   USSR launches *Sputnik,* world's first artificial satellite.

1958   *Explorer 1,* first U.S. satellite, successfully launched.

        (1 October) National Aeronautics and Space Administration (NASA) supercedes NACA.

        Space Task Group established at Langley to formulate Project Mercury, America's first manned spaceflight program.

        Jet Propulsion Laboratory (JPL) transferred to NASA.

        X-15, rocket-powered aircraft, makes its first powered flight; trio of X-15 planes fly through 1968.

1959   Atlas intercontinental ballistic missiles become operational.

1950s   During decade, NACA/NASA achievements include variable-sweep wings, lifting body and reentry studies, area rule concept in transonic aircraft.

1960   Army Ballistic Missile Agency, including von Braun team and Saturn booster project, transferred to NASA.

        Marshall Spaceflight Center (Alabama) formally dedicated.

1961   Goddard Spaceflight Center (Maryland) dedicated.

        (12 April) Yuri Gagarin (USSR) becomes first human being to orbit Earth.

Speaking before Congress, President John F. Kennedy commits America to a manned lunar landing.

Apollo program begins to take shape.

Manned Spacecraft Center begins with transfer of Space Task Group to Texas. Renamed Lyndon Baines Johnson Space Center in 1973.

Launch Operations Center formalized at Cape Canaveral (Florida). Renamed John F. Kennedy Space Center in 1963.

1962   John Glenn becomes first American to orbit Earth.

Telstar 1, the first transatlantic satellite television relay, is placed in orbit by NASA for AT&T.

Mariner 2, first satellite to fly by another planet, scans Venus, measures temperatures.

1963   Valentina Tereshkova (USSR) becomes first woman in space.

1965–
1966   Gemini series of two-man missions includes America's first "space walks."

1966   *Luna* (USSR) makes first hard landing of unmanned probe on lunar surface.

Surveyor, unmanned probe, makes successful soft landing on lunar surface.

Sergei Pavlovich Korolev, chief architect of USSR space program, dies.

1968   (21–25 December) *Apollo 8;* first manned spaceflight beyond Earth's gravitational pull, first manned circumlunar voyage.

1969   (16–20 July) *Apollo 11;* first successful manned mission to the moon; astronauts Neil Armstrong and Edwin Aldrin carry out experiments on lunar surface before rejoining Michael Collins in lunar orbit; the three return to Earth.

1960s   During the decade, continuing aeronautical work includes significant NASA flight research using large, supersonic aircraft, the XB-70.

1970   Airbus Industrie, European consortium, established to compete in global market for jet airliners.

1971   USSR launches *Salyut 1*, the first space station.

NASA's supercritical wing research program begun; yields significant benefits in transonic flight regimes.

1972   *Pioneer 10,* first spacecraft to survey outer planets, begins journey.

Apollo 17, final mission of Apollo/Saturn series for manned lunar landings.

1973   Skylab Program; three different American crews live and work in America's first space station.

1974   Wallops Flight Center dedicated (rocket flight test range; various designations since 1945).

Charles Lindbergh dies.

1975   European Space Agency (ESA) formally organized.

NASA's Aircraft Energy Efficiency Program initiates broad research into fuels, engines, composites, environmental engineering, and related projects for next-generation aircraft.

Apollo-Soyuz Test Project; spacecraft from U.S. and USSR rendezvous and dock in Earth orbit.

1976   Anglo-French Concorde, first supersonic airliner, enters service, followed by Tupolev Tu-144, the Soviet SST.

Viking 1 lander becomes first active scientific package to settle on Martian surface.

1977   First flight of NASA/Bell X-15 tilt-rotor aircraft.

Wernher von Braun dies.

1981   First flight of Space Shuttle with launch of *Columbia.*

1983   European Space Agency launches its first commercial satellite.

1984   President Ronald Reagan endorses plans for a permanent space station.

1985   Space Shuttle *Challenger* explodes after launch; seven crew members die.

Voyager, a composite aircraft, completes first unrefuelled, nonstop flight around the world.

1986   *Mir,* long-duration Soviet space station, is occupied.

1988   *Discovery* resumes launches of Space Shuttle series.

John C. Stennis Space Center dedicated (formerly Mississippi Test Facility and other designations).

1989   Berlin Wall demolished, symbolizing the dissolution of Soviet power and demise of cold war.

1980s	During the decade, NASA research verifies advantages of winglets; other studies benefit general aviation safety, air traffic control protocols, and cockpit architecture for airliners.
1992	Mergers of major aerospace companies in U.S. and abroad reshape the nature of global aerospace industries.
1994–95	First Russians fly aboard Space Shuttle; first U.S. astronauts carried to *Mir* aboard Soyuz booster; first Space Shuttle docking with *Mir*.
1997	Sojourner, remote-controlled module on Mars, captures imagination of the public.
1998	First major components of International Space Station carried into orbit.
1990s	New NASA initiatives for projects in SST studies, subsonic airliners, and general aviation technologies.
2000	(31 October) Expedition One (three-man crew) mission with first occupants of International Space Station.

Index

Page numbers in italics indicate illustrations.